TABLE OF CONTENTS

Math Enrichment

Introduction	2
Letter to Parents	4
Letter to Students	5
Student Progress Chart	6
Curriculum Correlation	8
Assessments	9

UNIT 1: Number
Base-Five System	13
Column Addition	14
Exchange-Rate Table	15
Keeping a Budget	16
Complement	17
Supplement	18
Decoding and Encoding	19
Advertising Table	20
Speed of Light	21
Casting Out 9's	22
Greatest Common Factor	23
Shortcut of Prime Factors	24

UNIT 2: Problem Solving
Estimation	25
Choosing the Operation	26
Identifying Extra/Needed Information	27
Solving Two-Step Problems	28
Choosing the Operation	29
Writing Equations	30
Interpreting the Quotient and the Remainder	31
Writing a Proportion	32
Using a Circle Graph	33
Using Broken-Line Graphs and Bar Graphs	34
Checking That the Solution Answers the Question	35
Using a Double-Line Graph	36

UNIT 3: Logic
Propositions	37
Riddles	38
Logic Tables	39
Riddles	40
Logic Tables	41
Possible Statements	42
Riddles	43
Logic Tables	44
Possible Combinations	45
Eliminating Choices	46

UNIT 4: Patterns
Estimating Sums	47
Increments	48
Arithmetic Sequences	49
Triangular and Square Numbers	50
Fibonacci Sequence	51
Geometric Sequences	52
Magic Square	53
Farey Series	54
Elements in a Set	55
Magic Squares	56
Inverse Proportion	57
Arithmetic and Geometric Sequences	58

UNIT 5: Measurement
Kilowatt-Hours	59
Precise Measurements	60
Lens Magnification	61
Determining Dates	62
Latitude and Longitude	63
Calories	64
Estimate Amount of Liquid	65
Other Units of Measure	66
Decibels	67
Karats	68
Interplanetary Weights	69
Personal Time Line	70

UNIT 6: Probability and Statistics
Sampling	71
Venn Diagrams	72
Venn Diagrams	73
Probability of Genotype	74
Batting Average	75
TV Ratings and Share	76
Venn Diagrams	77
Frequency	78
Mean	79
Odds	80
Odds	81
Probability	82

UNIT 7: Algebra and Geometry
Binomial Expression	83
Absolute Values	84
Nomograph	85
Networks	86
Helix	87
Fraction of Circumference	88
Pythagorean Theorem	89
Perimeter of Polygons	90
Circumference	91
Interior Angles	92
Volume of an Irregular Solid	93
ANSWER KEY	94

INTRODUCTION

Math Enrichment: GRADE 6

Students need to develop a solid sense of numbers and build the competence and confidence to compute, estimate, reason, and communicate to solve real-life problems. *Math Enrichment* engages students in the mathematics of real life–by reinforcing concepts, skills, and strategies that directly relate to daily living. Varied mathematical situations and intriguing content-area connections help students extend and enrich their understanding.

The activities in *Math Enrichment* complement and enhance your mathematics program. They spark both the teacher's and students' creativity and understanding. You will find this variety of activities engaging and challenging to *all* students!

ORGANIZATION

Math Enrichment is organized into seven units that focus on the essential areas of mathematics: number, problem solving, logic, patterns, measurement, probability and statistics, and algebra and geometry.

NUMBER
Includes different bases, money, codes, tables, shortcuts, and activities that reinforce addition, subtraction, multiplication, division, decimals, and fractions.

PROBLEM SOLVING
Includes tables, interesting facts, puzzles, graphs, and activities that reinforce the basic operations and estimation.

LOGIC
Includes stories, riddles, tables, diagrams, and activities that reinforce problem solving.

PATTERNS
Includes charts, stories, diagrams, and magic squares that reinforce skip-counting, multiplication, division, fractions, and negative numbers.

MEASUREMENT
Includes using telescopes and microscopes, calendars, maps, planets, and historical facts that reinforce multiplication, metric and standard measures, fractions, money, decimals, and time.

PROBABILITY AND STATISTICS
Includes surveys, tables, diagrams, codes, and graphs that reinforce categorizing, decimals, fractions, and probability.

ALGEBRA AND GEOMETRY
Includes area, negative numbers, fractions, angles, networks, perimeter, circumference, and solid figures.

INTRODUCTION
Math Enrichment: GRADE 6

USE

The activities in this book are designed for independent use by students who have had instruction in the specific skills covered in the lessons. Copies of the activity sheets can be given to individuals or pairs of students for completion. When students are familiar with the content of the worksheets, they can be assigned as homework.

To begin, determine the implementation that fits your students' needs and your classroom structure. The following plan suggests a format for this implementation.

1. **Administer** the Assessment Tests to establish baseline information on each student. These tests may also be used as post-tests when students have completed a unit.

2. **Explain** the purpose of the worksheets to the class.

3. **Review** the mechanics of how you want students to work with the activities. Do you want them to work in pairs? Are the activities for homework?

4. **Introduce** students to the process and purpose of the activities. Work with students when they have difficulty. Give them only a few pages at a time to avoid pressure.

ADDITIONAL NOTES

1. Parent Communication. Send the Letter to Parents home with students.

2. Student Communication. Encourage students to share the Letter to Students with their parents.

3. Bulletin Board. Display completed worksheets to show student progress.

4. Student Progress Chart. Duplicate the grid sheets found on pages 6-7. Record student names in the left column. Note date of completion of each lesson for each student.

5. Curriculum Correlation. This chart helps you with cross-curriculum lesson planning.

6. Have fun! Working with these activities can be fun as well as meaningful for you and your students.

Dear Parent:

During this school year, our class will be working with mathematical skills. We will be completing activity sheets that provide enrichment in the areas of number, problem solving, logic, patterns, measurement, probability and statistics, and algebra and geometry.

From time to time, I may send home activity sheets. To best help your child, please consider the following suggestions:

- *Provide a quiet place to work.*
- *Go over the directions together.*
- *Encourage your child to do his or her best.*
- *Check the lesson when it is complete.*
- *Go over your child's work, and note improvements as well as problems.*

Help your child maintain a positive attitude about mathematics. Let your child know that each lesson provides an opportunity to have fun and to learn. If your child expresses anxiety about these strategies, help him or her understand what causes the stress. Then talk about ways to eliminate math anxiety.

Above all, enjoy this time you spend with your child. He or she will feel your support, and skills will improve with each activity completed.

Thank you for your help!

Cordially,

Dear Student:

This year you will be working in many areas in mathematics. The activities are designed for fun and for real-life applications. You will complete puzzles and mazes, break codes, make tables and charts, and read maps. You will get to work with puzzles, graphs, planets, stories, maps, and money. These activities will show you fun ways to practice mathematics!

As you complete the worksheets, remember the following:

- *Read the directions carefully.*
- *Read each question carefully.*
- *Check your answers after you complete the activity.*

You will learn many ways to solve math problems. Have fun as you develop these skills!

Sincerely,

STUDENT PROGRESS CHART

STUDENT NAME	UNIT 1 NUMBER												UNIT 2 PROBLEM SOLVING												UNIT 3 LOGIC									UNIT 4 PATTERNS								
	13	14	15	16	17	18	19	20	21	22	23	24	25	26	27	28	29	30	31	32	33	34	35	36	37	38	39	40	41	42	43	44	45	46	47	48	49	50	51	52	53	54	55	56	57	58

STUDENT PROGRESS CHART

STUDENT NAME	UNIT 5 MEASUREMENT												UNIT 6 PROBABILITY AND STATISTICS												UNIT 7 ALGEBRA AND GEOMETRY											
	59	60	61	62	63	64	65	66	67	68	69	70	71	72	73	74	75	76	77	78	79	80	81	82	83	84	85	86	87	88	89	90	91	92	93

CURRICULUM CORRELATION

	Social Studies	Food and Nutrition	Music	Science	Recreation	Business
Unit 1: Number	14, 15, 16			21		16, 20
Unit 2: Problem Solving	27, 30, 32, 33, 34			35	32	25, 26, 28, 31, 34, 36
Unit 3: Logic	40, 43, 44	41, 44		42, 45	38, 39, 41	46
Unit 4: Patterns	54			48, 51	50, 57	48, 57
Unit 5: Measurement	62, 63, 66, 70			59, 61, 64, 65, 67, 68, 69	60	60
Unit 6: Probability & Statistics	80		77	71, 73, 74	72, 75, 77, 79	71, 76
Unit 7: Algebra & Geometry	89			87, 88	91	

Name _____ Date _____

ASSESSMENTS

Assessment: Units 1 and 2

1. What is the greatest common factor of 60, 156, and 378?
    a. 6
    b. 10
    c. 15
    d. 60

2. What is the product using complements of the factors?
   86 × 93 =
    a. 98   b. 179   c. 7,998   d. 74,958

3. The school bond issue got 8,011 votes of the 8,541 votes cast. How many people voted against the school bond issue?
   What equation would you use to solve the problem?
    a. 8,011 + 8,541 = n
    b. 8,541 - 8,011 = n
    c. 8,541 × 8,011 = n
    d. 8,541 ÷ 8,011 = n

4. **The Inner Planets**

Planet	Surface Area	Maximum Orbital Distance from the Sun
Mercury	74,800,000 km²	69,800,000 km
Venus	460,270,000 km²	109,000,000 km
Earth	510,070,000 km²	152,000,000 km
Mars	55,870,000 km²	249,000,000 km

Mercury's maximum orbit is what percent of Mars's maximum orbit?
    a. 28%   b. 72%   c. 50%   d. 75%

Name _____ Date _____

ASSESSMENTS

Assessment: Units 3 and 4

1. Use the logic table to help you answer the question.

	Lisa	Annie	Darius	Brianna
Soccer				
Skateboard				
Tennis				
Swim				

   A. Darius, Brianna, Annie, and Lisa like to play soccer, ride skateboards, play tennis, and swim. Among these hobbies, each person has a favorite.
   B. Darius likes activities with wheels.
   C. Lisa and Brianna do not like to get water in their ears.
   D. The person who likes to play tennis is a girl and is friends with both Annie and Brianna.
   E. Lisa, Annie, and Darius like to watch soccer games.

   Who likes to swim?
   a. Lisa **b.** Annie **c.** Darius **d.** Brianna

2. What is the sum of this magic square?

15	50	43
64	36	8
29	22	57

   a. 108 **b.** 288 **c.** 324 **d.** 32,250

3. Which of the following is an arithmetic sequence?
   a. 2.5, 2.5, 2.5, 2.5, . . .
   b. 1.3 + 1.2, 1.3 + 1.2 + 1.3, 1.3 + 1.2 + 1.3 + 1.2, . . .
   c. 25, 22, 29, 26, 23, . . .
   d. 2.4, 3.6, 3.9, 5.1, 5.4, . . .

4. C = {2, 4, 6, 8, 10, 12, 14}
   What is C_4?
   a. 4 **b.** 8 **c.** 2, 4, 6, 8 **d.** None of these

Name _____ Date _____

ASSESSMENTS

Assessment: Units 5 and 6

1. Order the measurements from the most precise to the least precise:
   2,130 km; 2,000 km; 2,132 km; 2,100 km

   a. 2,132 km; 2,130 km; 2,100 km; 2,000 km

   b. 2,000 km; 2,130 km; 2,132 km; 2,100 km

   c. 2,000 km; 2,100 km; 2,130 km; 2,132 km

   d. not enough information given

	Mercury	Venus	Earth	Moon	Jupiter	Saturn	Neptune
1 lb	0.28	0.85	1.0	0.16	2.6	1.2	1.4

2. Use the equivalence chart for interplanetary weights to solve the problem. Tran stopped in on the moon to purchase cheese. He wants to buy 1 lb of cheese. How much will he have the clerk weigh out on the moon?
   a. 1.16 lb   b. 1 lb   c. 0.16 lb   d. .84 lb

3. The Jacoby family likes to listen to CDs. They own 40 CDs, of which 10 are classical music, 4 are rock and roll, 6 are rap, 2 are movie soundtracks, 15 are folk music, and 3 are jazz. Everyone likes to listen to classical music, but Matthew doesn't like rock and roll, the parents don't like rap music, and Kate doesn't listen to folk music. Matthew is the only one who listens to jazz, and Kate is the only one who listens to movie soundtracks. The information is illustrated on this Venn diagram.

   How many CDs does Matthew listen to, altogether?
   a. 3   b. 18   c. 25   d. 34

4. What are the odds that 2 people in a class of 30 would have the same birthday?

   a. There is a 15% chance that everyone has a different birthday.

   b. There is a 71% chance that at least two people have the same birthday.

   c. There is a 29% chance that at least two people have the same birthday.

   d. There is no chance that at least two people have the same birthday.

Assessment: Unit 7

1. Solve.

   $(^-3 \times {}^-4) \times (^+8 \times {}^-1) =$

   a. $^+96$

   b. $^+16$

   c. $^-96$

   d. $^-56$

   Network A

   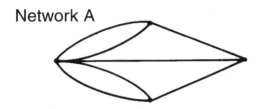

2. How many vertices of an even degree does Network A have?

   a. 0   **b.** 1   **c.** 2   **d.** 3

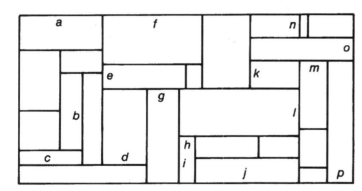

a = 11	i = 6
b = 10	j = 14
c = 8	k = 4
d = 5	l = 6
e = 4	m = 3
f = 12	n = 3
g = 4	o = 3
h = 2	p = 3

3. What is the length of $k + o$?

   a. 7   **b.** 11   **c.** 12   **d.** 14

4. What is the perimeter of the large rectangle in number 3?

   a. 50   **b.** 88   **c.** 128   **d.** none of these

Name _____ Date _____

NUMBER

Unit 1: Base-Five System

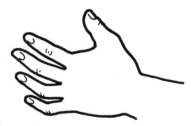

In base ten, there are place values of 1; 10; 100; 1,000; and so on. The numbers 0, 1, 2, 3, 4, 5, 6, 7, 8, and 9 are used to fill these place values. The base-five system has place values of 1, 5, 25, 125, 625, and so on. The numbers 0, 1, 2, 3, and 4 are used to fill these place values.

hundreds	tens	ones
2	2	7

$227_{10} = 1,402_5$

625's	125's	25's	5's	1's
	1	4	0	2

To convert a number from base five to base ten, simply find the number of units in the different place-value columns and add them up.

625's	125's	25's	5's	1's
1	2	1	2	3

$12,123_5 = 625 + 125 + 125 + 25 + 5 + 5 + 3 = 913_{10}$

Write the base-ten equivalent of each number.

1. 23_5 _____
2. 341_5 _____
3. 120_5 _____
4. 221_5 _____

To convert a number from base ten to base five:

a. Start with the largest place value in the base-five system that is smaller than the number. Write a digit that represents the number of times that the place value can be added to itself and still be smaller than the number.

b. Subtract that place-value sum from the base-ten number.

c. If the next-lowest place value is smaller than the difference, repeat steps **a** to **c**. If it is greater, write a 0 and repeat this step.

d. Continue this process until the remainder is 0.

3,125's	625's	125's	25's	5's
	2			

$1,380 - 1,250 = 130$

3,125's	625's	125's	25's	5's	1's
	2	1			

$1,380_{10} = 21,010_5$

Write the base-five equivalent of the base-ten number.

5. 104 _____
6. 71 _____
7. 200 _____
8. 324 _____

NUMBER

Column Addition

Most people in the world today add columns of numbers with the method you have been taught. But throughout history, different methods have been used. One method is **retrograde**, or **inverse**, **addition**.

a. Start at the left. Add 5 + 3. Write 8 below the line.
b. Add 5 + 8 + 2 = 15. Write 5 below the line. Add 1 to the 8 from step a. Cross out the 8 and write 9 below.
c. Add 3 + 1 + 2 = 6. Write 6 below the line.
d. Add 9 + 6 + 8 = 23. Write 3 below the line. Add 2 to the 6 from step c. Cross out the 6 and write 8 below.
e. The numbers under the line that are not crossed out form the answer: 9,583.

```
      5,539
      3,816
    +   228
      8,563
      9 8
      9,583
```

Use inverse addition. Show all the steps.

1.	7,654	2.	1,732	3.	2,747	4.	661	5.	4,804
	1,483		71,654		812		1,004		2,402
	+ 772		+ 3,994		+ 39		+ 382		+ 1,206

In the sixteenth century, the Dutch mathematician Gemma Frisius devised another method of addition. Using his system, you add partial sums in the same way that you add partial products in multiplication.

```
       753
       429
     + 202
        14
         7
        13
     1,384
```

Add. Use Gemma Frisius's method. Show all the steps.

6.	209	7.	235	8.	378	9.	168	10.	1,333
	830		697		816		725		7,007
	+ 326		+ 198		+ 22		+ 264		+ 685

NUMBER

Exchange-Rate Table

In the United States, our currency is the U.S. *dollar.* Italian currency is called the *lire,* British currency is called the *pound,* Swiss currency is called the *franc,* and German currency is called the *mark.* To find out how much the U.S. dollar is worth in another country's currency, you must find the exchange rate. For instance, if the Italian exchange rate is 2,000 lire to one U.S. dollar, that means that one U.S. dollar is worth 2,000 lire. You can find the exchange rate in the newspaper.

1 U.S. dollar =
    3.5 German marks
    2,000 Italian lire
    .5 British pound
    2.5 Swiss francs

Your family decides to take a trip to Europe. You will visit England, Germany, Switzerland, and Italy. You are given $100.00 spending money and decide to spend about $25.00 in each country.

Use the exchange-rate table to solve.

1. How many British pounds do you receive for $25.00? _____

2. How many German marks do you receive for $25.00? _____

3. You buy a statuette of the Leaning Tower of Pisa in Italy and a stuffed bear in England. The statuette costs 9,000 lire and the bear costs 5 pounds. How much does each item cost in U.S. dollars? _____

4. In England you see a cassette tape that sells for 3.50 pounds. You have seen the same cassette in the United States for $5.50. Is it less expensive in the United States or in England? _____

5. You want to buy postcards in Germany. If each postcard costs 1 mark, and you spend $3.00, how many postcards can you buy? _____

6. In the airport, you exchange 8 German marks for Swiss francs. How many francs do you receive? _____

7. In Switzerland, your father buys a watch for 375 francs. What is its price in U.S. currency? _____

8. When you leave Switzerland you have 10 francs left. You want to exchange the Swiss francs for Italian lire. How many lire will you receive? _____

9. You and your brother go out for lunch in Italy. The bill comes to 20,500 lire. How much is that in U.S dollars? _____

10. When you arrive home you still have 5,600 lire and 15 francs left. How much is that in U.S. currency? _____

Name _____   Date _____

NUMBER

Keeping a Budget

Latoya wants to buy a new bicycle. She decides to make a budget to help her keep track of her money.

Every day Latoya walks 6 dogs in her neighborhood for $0.50 each. Every week, she sees two movies for $3.50 each. She also buys a book for $6.50 every week. Every school day she spends $0.90 on lunch. Her budget looks like this.

Weekly Budget

Income	Expenses
Walking Dogs-$21.00	$7.00-Movies
	$6.50-Book
	$4.50-Lunch
Total Income-$21.00	$18.00-Total Expenses

Solve.

1. If Latoya saves the money she has left at the end of each week, how many weeks will it take her to save enough money to buy a $75.00 bicycle? _____

2. If Latoya were to start walking 7 dogs per day, how many weeks would it take her to save for the bicycle? _____

Latoya figures that she can live on less money than she is now spending each week. So, she makes a new budget which allows only 1 movie per week, and no books. However, she will allow herself $2.00 a week for popcorn. On this new budget:

3. How much will Latoya save each week? _____

4. How much more money is she saving than on her original budget each week? _____

5. How many weeks will it take Latoya to save enough money to buy the new bicycle? _____

6. If Latoya stops walking one of the 6 dogs, how will it affect her weekly savings? _____

NUMBER

Complement

The **complement** of a number is the amount by which that number is less than 100. The complement of 97 is 3 (100 - 3 = 97). The complement of 95 is 5. You can use complements to help you find the product of two 2-digit numbers.

97 × 95 = ☐

a. Subtract the complement of one of the numbers from the other number.

97 - 5 = 92   OR   95 - 3 = 92

These are the first two digits of the final product.

b. Multiply the complements of the two numbers together.

5 × 3 = 15

These are the last two digits of the final product.
c. Combine the two numbers to get your answer.
92,15   97 × 95 = 9,215

Find the product using the complements of the factors.

1. 74 × 97 = _____   **2.** 94 × 92 = _____

3. 96 × 83 = _____   **4.** 99 × 78 = _____

5. 89 × 98 = _____   **6.** 92 × 88 = _____

7. 94 × 84 = _____   **8.** 98 × 76 = _____

9. 95 × 97 = _____   **10.** 96 × 89 = _____

11. 97 × 68 = _____   **12.** 82 × 95 = _____

13. 88 × 94 = _____   **14.** 96 × 76 = _____

15. 92 × 93 = _____   **16.** 84 × 95 = _____

17. 93 × 87 = _____   **18.** 96 × 96 = _____

19. This shortcut to multiplication only works under one condition. What is that condition? (HINT: All of the exercises on this page conform to that condition.)

_____

NUMBER

Supplement

The **supplement** of a number is the amount by which that number is greater than 100. The supplement of 104 is 4 (100 + 4 = 104). The supplement of 103 is 3. You can use supplements to help you find the product of two 3-digit numbers.

104 x 103 = ▢

a. Add one of the numbers to the supplement of the other number.

104 + 3 = 107   **OR**   103 + 4 = 107

These are the first three digits of the final product.

b. Multiply the supplements of the two numbers together.

3 x 4 = 12

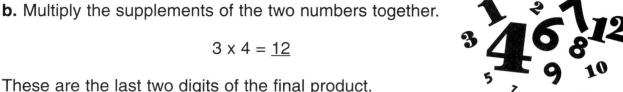

These are the last two digits of the final product.

c. Combine the two numbers to get your answer.

104 x 103 = 10,712

Find the product using the supplements of the factors.

1. 115 x 105 = _____   **2.** 134 x 102 = _____

3. 114 x 104 = _____   **4.** 120 x 102 = _____

5. 121 x 103 = _____   **6.** 105 x 110 = _____

7. 118 x 105 = _____   **8.** 103 x 129 = _____

9. 125 x 101 = _____   **10.** 110 x 109 = _____

11. 108 x 108 = _____   **12.** 120 x 104 = _____

13. 103 x 117 = _____   **14.** 114 x 102 = _____

15. This shortcut to multiplication only works if the product of the supplements is less than 100. If the product of the supplements were greater than 100, how would you change step **c,** above, so that you could arrive at the correct answer?

_____

_____

Name _____ Date _____

NUMBER

Decoding and Encoding

14 26 7 19 25 6 18 15 23 8
__ __ __ __ __ __ __ __ __ __

19 22 26 15 7 19 2 14 18 13 23 8
__ __ __ __ __ __ __ __ __ __ __ __

These numbers are really a sentence written in **code.** Each number represents a letter in the alphabet. In this code, A = 26, B = 25, C = 24, and so on to Z = 1. To translate the message, simply match the number to its corresponding letter. This process is called **decoding** or **deciphering.** When you translate a message into code it is called **encoding** or **enciphering.**

1. Decode the message at the top of the page.

2. Encode the following message.
There is a spy in the bookkeeping department.

_____

_____

_____

3. Decode the following message.

13 12   7 19 22 8 11 2 18 8
__ __ , __ __ __ __ __ __ __ __

14 2 23 12 20 21 15 6 21 21 2
__ __ __ __ __ __ __ __ __ __ __

4. Codes are a basic part of your life. Every day, you see and use codes that are all around you. List five codes that you use or see.

_____

_____

_____

NUMBER

Advertising Table

Advertisers use television commercials to convince you to use their product or service. But television commercials are expensive. A 30-second commercial, or *spot*, on *Those Blinker Kids*, a prime-time comedy series, costs $500,000. A 60-second spot costs 1.8 times as much as a 30-second spot, but it is twice as effective. The effectiveness of a television commercial can be rated on a scale from 0 to 100. The higher the number, the more people are likely to respond to the commercial. The table shows the cost and effectiveness of 30- and 60-second spots on several different shows.

1. Complete the table.

		Those Blinker Kids	Tommy House Talk Show	Life's Like That!	Early Morning	Late P.M.
Cost	30	$500,000	$45,000	$20,000	$17,000	$7,500
	60	$900,000	$81,000	$36,000	$30,600	$13,500
Effectiveness	30	50	40	16	13.6	6
	60	100	80	32	27.2	12

2. Which are more effective, ten 30-second spots on *Early Morning* or eleven 60-second spots on *Late P.M.*?

Ten 30-second spots on Early Morning (136 vs. 132)

3. Which costs more?

Ten 30-second spots on Early Morning ($170,000 vs. $148,500)

You can figure out how economical a spot is by dividing its cost by its effectiveness rating. The lower the quotient, the more economical the spot is.

4. Complete the chart.

		Those Blinker Kids	Tommy House Talk Show	Life's Like That!	Early Morning	Late P.M.
Economy	30	10,000	1,125	1,250	1,250	1,250
	60	9,000	1,012.5	1,125	1,125	1,125

Unit 1: Number

NUMBER

Speed of Light

Have you ever watched someone a distance away from you bounce a ball? You see the ball bounce before you hear it bounce. That's because light travels much faster than sound. The light that reflects off the ball reaches your eyes before the sound that the ball makes reaches your ears. In fact, it is theorized that nothing can travel faster than light, which has a speed of 186,000 miles per second.

1. If the sun's light takes 8 minutes and 20 seconds to reach Earth, how far away is the sun? _____

2. Imagine that you are on the moon when it eclipses Earth. How would the amount of time it takes sunlight to reach you differ from what you are used to on Earth?

   _____

3. The moon is 250,000 miles from Earth. How much time, to the nearest second, would it take sunlight to reach you there during the solar eclipse? _____

4. Assume that the moon and Earth are the same distance from the sun. How long, to the nearest second, does it take for sunlight to reach the moon and reflect to Earth?

   _____

Sometimes light reflects off Earth and hits the moon. This is called *Earthlight*. Earthlight enables you to see the part of the moon that is not lit by the sun.

5. How long, to the nearest second, does it take Earthlight to reach Earth from its source? _____

6. Imagine that you are on a spaceship traveling away from the sun at 186,005 miles per second. If you look out the rear window, will you see sunlight? Explain.

   _____

Name _____ Date _____

NUMBER
Casting Out 9's

Here's a different way to check your addition. It's a method called casting out 9's. To do this check of your addition, simply add the digits of your addends until there is only one digit left. If the number is less than 9, that number is your digit sum. If the sum is 9, the digit sum is 0. Here are some examples.

Did You Know?

	Step 1	Step 2	Step 3
582 + 272 854	5 + 8 + 2 = 15 2 + 7 + 2 = 11	1 + 5 = 6 1 + 1 = 2	2 + 6 = 8

This number should equal the digit sum of the correct answer.
8 + 5 + 4 = 17; 1 + 7 = 8

Solve. Ring the number of any exercise with an incorrect solution.

1.
   13,907
+ 2,440
  16,347

final digit sum: _____

digit sum: _____

2.
   13,791
   97,654
+   111
 110,556

final digit sum: _____

digit sum: _____

3.
  18
  28
+ 46
  92

final digit sum: _____

digit sum: _____

4.
  64,002
180,222
+   101
244,325

final digit sum: _____

digit sum: _____

5.
  33,330
  27,277
+ 11,881
  72,488

final digit sum: _____

digit sum: _____

6.
192,864
 14,001
+ 1,221
218,086

final digit sum _____

digit sum: _____

7.
   8,341
  55,442
+ 277,280
 331,063

final digit sum: _____

digit sum: _____

8.
624,098
    28
+ 86,180
720,206

final digit sum: _____

digit sum: _____

9. Can you think of any disadvantage of using casting out 9's as a method of checking your addition?

Name _____ Date _____

NUMBER

Greatest Common Factor

You can find the **greatest common factor** of two numbers quickly by using repeated division. For example, to find the greatest common factor of 160 and 358:
a. Divide the larger number by the smaller, writing the remainder as a whole number.
b. Then divide the divisor by the remainder.
a. Repeat step **b** until a remainder of 0 is reached.

$358 \div 160 = 2$ R38

$160 \div 38 = 4$ R8
$38 \div 8 = 4$ R6
$8 \div 6 = 1$ R2
$6 \div \underline{2} = 3$ R0

The divisor that yields a remainder of 0, which in this case is 2, is the greatest common factor of the two numbers.

Use this method to find the greatest common factor.

1. 88 and 242 _____   **2.** 145 and 270 _____   **3.** 144 and 300 _____

The same method can be used to find the greatest common factor of three numbers. For the numbers 60, 156, and 378, first find the greatest common factor of any two, for example 60 and 156.

$156 \div 60 = 2$ R36 $60 \div 36 = 1$ R24
$36 \div 24 = 1$ R12 $24 \div \underline{12} = 2$ R0

Then take the greatest common factor, 12, and find the greatest common factor of it and the remaining number, 378.

$378 \div 12 = 31$ R6 $12 \div \underline{6} = 2$ R0

Then find the greatest common factor of the two, 12 and 6.

$12 \div \underline{6} = 2$ R0

The greatest common factor of 60, 156, and 378 is 6.

Write the greatest common factor.

4. 30, 225, 375 _____   **5.** 32, 128, 400 _____   **6.** 180, 240, 72 _____

7. 56, 72, 332 _____   **8.** 75, 60, 275 _____   **9.** 144, 300, 30 _____

Name _____ Date _____

NUMBER

Shortcut of Prime Factors

Multiplying fractions with large numerators or denominators can be simpler than it seems, if you use a **shortcut of prime factors**. This shortcut involves writing each numerator and denominator as a product of prime factors. To multiply $\frac{9}{22} \times \frac{8}{25} \times \frac{11}{18}$, first find the prime factors of each number.

$$\frac{9}{22} = \frac{3 \times 3}{2 \times 11} \qquad \frac{8}{25} = \frac{2 \times 2 \times 2}{5 \times 5} \qquad \frac{11}{18} = \frac{11}{2 \times 3 \times 3}$$

Then write the product as one fraction.

$$\frac{9}{22} \times \frac{8}{25} \times \frac{11}{18} = \frac{3 \times 3 \times 2 \times 2 \times 2 \times 11}{2 \times 11 \times 5 \times 5 \times 2 \times 3 \times 3}$$

Divide both the numerator and the denominator by each common factor, and multiply the factors.

$$\frac{\cancel{3} \times \cancel{3} \times \cancel{2} \times \cancel{2} \times 2 \times \cancel{11}}{\cancel{2} \times \cancel{11} \times 5 \times 5 \times \cancel{2} \times \cancel{3} \times \cancel{3}} = \frac{1 \times 1 \times 1 \times 1 \times 2 \times 1}{1 \times 1 \times 5 \times 5 \times 1 \times 1 \times 1} = \frac{2}{25}$$

Find each product using the shortcut of prime factors. Show all work.

1. $\frac{20}{21} \times \frac{13}{15} \times \frac{7}{26} =$

2. $\frac{9}{13} \times \frac{26}{27} \times \frac{7}{26} =$

3. $\frac{6}{7} \times \frac{14}{15} \times \frac{5}{11} =$

4. $\frac{5}{9} \times \frac{10}{11} \times \frac{22}{25} =$

5. $\frac{15}{16} \times \frac{24}{25} =$

6. $\frac{13}{18} \times \frac{4}{7} \times \frac{9}{26} =$

PROBLEM SOLVING

Unit 2: Estimation

The TV game show *Shop Around the Clock* gives its contestants a chance to shop for prizes. In order to keep their purchases, the contestants must not spend more than the allotted amount of either time or money.

PRIZES AND PRICES
Contestants may select unlimited amounts of items.

Can of noodle soup	$0.59	Silver tiepin	$12.29
Bowling ball	$31.40	Four-day vacation	$12,432.00
Computer	$1,429.89	Blanket	$18.39
Record album	$7.69	Dictionary	$12.29
Pan	$14.56	Year's supply of peanuts	$139.68
Camera	$52.45	Rowboat	$1,318.18
Box of detergent	$8.38	Miniature car	$3,569.22
Emerald ring	$423.78		

List what each contestant can purchase within the limit of time and money. Estimate unless you are very close to the amount allowed for purchases.

1. Marsha has $129.28. She has 1 min.

2. Larry has $24,567.88. He has 1 min.

3. Andrew has $867.45. He has 1 min.

4. Penny has $34,298.00. She has 1 min.

5. Carla has $10.05. She has 15 s. If she spends exactly that amount, she will win $1,000.00.

6. Alice has $5.90. She has 5 s. If she spends exactly that amount, she will win $2,000.00.

Name _____ Date _____

PROBLEM SOLVING

Choosing the Operation

Using the information below, fill in the correct amounts in the savings-account book. Figure out the final balance. Then answer the questions.

SAVINGS-ACCOUNT PASSBOOK

Date	Deposit	Withdrawal	Balance

Jamie has a job delivering papers. She makes $25 per week plus tips. She decides to start a savings account to manage her money. During the first week of her job, Jamie earns $15 in tips. She deposits $30 on September 15. During the next week, she makes $12 in tips. On September 22, she deposits $27. During the third week of her job, she makes a total of $42. She saves $30 and deposits it on September 29. On November 3, Jamie withdraws $43 to buy a present. During the next week, she earns a total of $40, but she has to give her teacher a $50 down payment toward a class trip. On November 11, she withdraws the rest of the money she needs for the down payment. On November 17, she deposits $25 from her most recent paycheck.

1. How much did Jamie earn altogether during the first week? _____

2. How much did she earn altogether during the second week? _____

3. How much did Jamie earn in tips during the third week? _____

4. How much money did Jamie have left after her first deposit? _____

5. Could Jamie afford to spend $8 on stockings after she received her second paycheck? _____

6. How much money did Jamie earn but not deposit during the first three weeks of her job? _____

© Steck-Vaughn Company

26

Unit 2: Problem Solving
Math Enrichment 6, SV8397-8

PROBLEM SOLVING

Identifying Extra/Needed Information

Solve. If possible, find the missing information from outside sources.

The first skyscraper was built in Chicago in 1885. Chicago now has three of the tallest skyscrapers, including the 110-story Sears Tower. The other two are the 1,136 ft Standard Oil building and the John Hancock Center, which is 1,107 ft tall (to the roof) and also has a 344 ft spire. New York City has five of the tallest buildings in the world. The most famous are the 1,350 ft Twin Towers of the World Trade Center and the 102-story Empire State Building, which has a 222-ft antenna on its roof. Two more of the tallest buildings are the 75-story Texas Commerce Tower in Houston, which is 1,002 ft tall, and Toronto's First Bank Tower, three stories shorter than the Texas Commerce Tower.

1. What is the difference in height between the Sears Tower and the second tallest building in Chicago?

2. How many feet are there between the roof of the Sears Tower and its highest point?

3. How many stories taller is the Sears Tower than the first skyscraper?

4. Is the Empire State Building twice as tall as the 607 ft tall Space Needle in Seattle?

5. How much shorter is the 935 ft First Bank Tower than the Texas Commerce Tower?

6. How many stories are there altogether in the Twin Towers?

7. How many feet tall would the three tallest buildings in Chicago be if added together?

8. How many stories taller are the Twin Towers of the World Trade Center than the First Bank Tower in Toronto?

9. What is the difference in total height between the three tallest buildings in Chicago and the five tallest buildings in New York?

10. What is the total height of all the skyscrapers in New York City?

PROBLEM SOLVING

Solving Two-Step Problems

Hern's Department Store is having a storewide sale. Use the information in the following advertisement to answer the questions below.

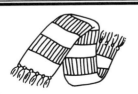

Hern's One-Week Sale!
Big Saving!

Gloves:	$8.75	Hats:	$9.50
Scarves:	$7.50	Stockings:	$3.99/pr.
Corduroy pants:	$17/pr.	Ties:	$11
Wool skirts:	$22	Umbrellas:	$14
Flannel shirts	$12	Socks:	regularly $2.25/pr., now 5 for $10

1. How much will it cost Candy to buy a pair of gloves, a scarf, and two wool skirts? _____

2. Cindy plans to buy 10 pairs of socks. How much will she save over the regular price? _____

3. How much will it cost Maria to buy a tie for each of her three sons and a hat for her husband? _____

4. Jim returns a $33.99 jacket for credit. He buys three pairs of corduroy pants and uses the credit to pay for a part of the purchase price. How much did he have to pay in addition to his store credit? _____

5. Kim will buy an umbrella for herself and one for her sister. Her mother also wants an umbrella and a pair of gloves and gave Kim $25 for her purchases. How much is the total cost for all the items? Did Kim's mother give Kim enough money to buy the items she asked for? _____

6. Emily is buying three skirts as presents. She has $80. What else can she buy? _____

Name _____ Date _____

PROBLEM SOLVING

Choosing the Operation

Read the paragraph and problems below. Write the letter of the correct operation after each problem (A = addition; S = subtraction; M = multiplication; D = division). Then solve the problem.

John and his four friends are going to the beach. It takes two hours to drive the 65 miles to get there. Gas is $1.20 per gallon, and John's car gets 25 miles per gallon. They drive over a bridge that has a $1.00 toll (each way). Parking at the beach costs $2.25 per hour or per fraction of an hour. At the beach, they rent an umbrella for $5.00. After $4\frac{1}{2}$ hours, they leave the beach and go home.

1. How long did the trip take?

_____

2. Mary and Phil split the parking fee. How much did each pay?

_____

3. Rich and Cathy pay for gas and tolls. How much did each pay?

_____

4. John paid to rent the umbrella. Who paid the most on the trip? Who paid the least? What is the amount of the difference between the two?

_____

5. It costs $7.00 per person for a round-trip bus ride to the beach. Would it cost more or less for all of them to take the bus rather than drive to the beach? By how much?

_____

6. The bus averages about 26 miles an hour on the way to the beach. Does it take more or less time to take the bus than it does to drive? How long does the trip to the beach by bus take?

_____

© Steck-Vaughn Company

Unit 2: Problem Solving
Math Enrichment 6, SV8397-8

PROBLEM SOLVING

Writing Equations

Write the equation you would use to solve each problem. Then write the equations and their solutions in the puzzle.

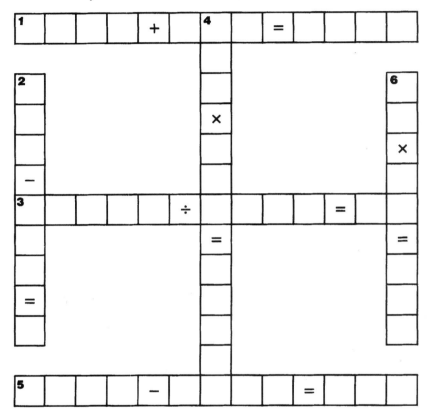

1. Inham had 8,567 registered voters a week before the election. That week, 935 people register. How many voters did Inham have for the election?

2. Mr. Jay went door to door through his town asking people to sign a petition. He asked 799 people; 792 signed it. How many people didn't sign it?

3. The fund drive raised $71,250 from 2,850 people. Everyone gave the same amount. How much was each contribution?

4. A traveling salesman sells vacuum cleaners for $222. One day, he sold 320 vacuums. How much money did he receive?

5. The mayor of Amber got 8,011 votes of 8,541 votes cast. How many people didn't vote for her reelection?

6. Each of 45 census takers is given a list of 15 houses to survey. How many houses in all will they survey?

Name _____ Date _____

PROBLEM SOLVING

Interpreting the Quotient and the Remainder

Jim and Audrey own a catering hall. They can cater parties for 20 to 200 people. Their hall is used for birthday parties, weddings, retirements, and other events. The chart below lists some of their available supplies. Use it to answer the following questions.

Tables	Number of people	Serving dishes	Number of servings	Silverware	Per Box	Napkins	Per box
Round	6	3 quart	5	Forks	16	Sm. box	100
Square	4	5 quart	8	Spoons	16	Med. Box	250
Rectangular	10	8 quart	12	Knives	24	Lg. Box	500
Oblong	12	10 quart	20				
Octagon	8	15 quart	30				

1. Audrey plans a birthday party for 160 people. What is the least number of tables that can be set up to accommodate 160 guests?

2. If Audrey wanted everyone to be seated at the same size table, which type of table and how many of them would she need?

3. Each person coming to the party will need 2 forks, 2 spoons, and 2 knives. How many complete boxes of utensils will be used?

4. A medium-size box of napkins has been opened and is $\frac{1}{5}$ full. How many more boxes will have to be opened for the party?

5. At a wedding party, soup will be served to 136 guests. Which serving dishes and how many of each should be used to serve exactly 136 guests?

6. If 8 people sit at one rectangular table, how many round tables would be needed to seat the rest of the wedding guests?

7. Guests at the wedding party have a choice of salad or vegetable. To be sure there is enough, Audrey makes enough portions of each to serve $\frac{3}{4}$ of the guests. Which serving dishes and how many of each will she use for each choice?

8. For each place setting at the wedding two forks, a spoon, and a knife are used. How many boxes need to be opened?

© Steck-Vaughn Company

Unit 2: Problem Solving
Math Enrichment 6, SV8397-8

PROBLEM SOLVING

Writing a Proportion

Edmunton took a survey of its citizens to determine which facilities should be included in its new park. Use the results of the survey below to answer each question.

	Adults		Teenagers		Senior citizens		Children	
Number polled	157		80		56		25	
	Yes/no		Yes/no		Yes/no		Yes/no	
Playground	79	77	41	39	14	42	16	9
Tennis courts	92	65	36	44	19	37	7	18
Pool	77	69	52	28	27	29	15	10
Skating rink	80	76	32	48	18	38	10	15
Garden	76	81	23	57	44	12	5	20

1. What is the ratio of adults who want to have a pool to adults who don't want to have a pool?

2. What is the ratio of adults who don't want to have a playground to senior citizens who don't want to have a playground?

3. Is the proportion of adults in favor of building a pool equal to the proportion of teenagers in favor of building a pool?

4. Is the proportion of children in favor of building a skating rink equal to the proportion of teenagers in favor of building a skating rink?

5. Is the proportion of children in favor of building a skating rink equal to the children not in favor of building a pool? Would a *yes* answer mean they are the same children?

6. What is the ratio of people in favor of building tennis courts to the number of people opposed to building tennis courts?

7. What is the ratio of senior citizens in favor of having a garden to the total number of people polled?

8. Assuming the survey is representative of the total population, about how many senior citizens in the total population don't want to have a skating rink?

Name _____ Date _____

PROBLEM SOLVING

Using a Circle Graph

One of the things archeologists hope to learn when they discover a site is the kinds of crops prehistoric farmers grew. One archeologist excavated a site at Snaketown, Arizona. He found seeds for mesquite, sahuaro, spiral beans, and other types of beans, all used by ancient Hohokam farmers. He found that spiral beans accounted for 5 of every 100 seeds. He also found that the farmers grew 5 times more mesquite than spiral beans and $\frac{1}{5}$ more sahuaro than mesquite. Bean-like seeds accounted for $7\frac{2}{5}$ times the number of spiral-bean seeds. The remaining seeds were too few to have separate categories, and were labeled *Other*.

SNAKETOWN, ARIZONA SEED QUANTITIES
(per 100 seeds excavated)

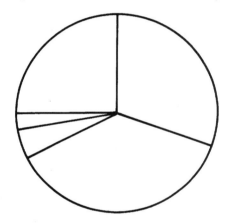

Use the information in the paragraph to complete the graph by filling in the percent of spiral bean, sahuaro, mesquite, beanlike, and other seeds found.

1. What would the spiral-bean percentage be if the site had yielded $2\frac{1}{2}$ times as many spiral-bean seeds?

_____

2. What would the percent of sahuaro seeds found be if only $\frac{1}{3}$ the number of these seeds had been unearthed?

_____

3. If $\frac{1}{10}$ less mesquite seeds had been found, what would the new percent for these seeds be?

_____

4. What is the ratio of the number of mesquite seeds found to the number of sahuaro seeds?

_____

Name _____ Date _____

PROBLEM SOLVING

Using Broken-Line Graphs and Bar Graphs

Photolink, Inc., sells photographs of exotic places around the world to newspapers, magazines, and publishers. In 1985, they sold 800 photos of the North Pole and 450 photos of the Taj Mahal. Photographs of Tahiti and the Grand Canyon each account for 780 photographs. Some other areas photographed are Australia's Great Barrier Reef and the Great Wall of China. Yearly sales are divided into quarters. First-quarter sales amounted to 900 photos. The second-quarter sales amounted to 960, and third-quarter sales amounted to 870 photographs. Fourth-quarter sales were $\frac{1}{3}$ greater than were first-quarter sales.

Use the information to complete the bar graph and the broken-line graph. Then solve each problem.

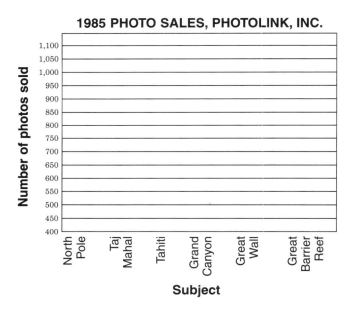

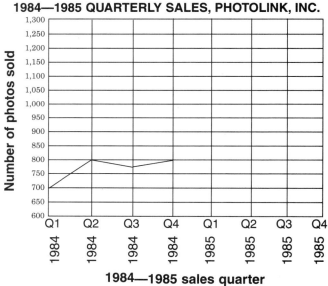

1. Which was the best-selling photograph of 1985? How many were sold? _____

2. What was the general sales trend for Photolink, Inc., in 1985? _____

3. There were 660 Barrier Reef photos sold and 1,100 Great Wall photos sold. Now, which photos are the first, second, and third most popular? _____

Name _____ Date _____

PROBLEM SOLVING

Checking That the Solution Answers the Question

Of the nine known planets orbiting our sun, the first four are called the inner planets. The inner planets and their maximum orbital distances from the sun are Mercury at 69,800,000 km, Venus at 109,000,000 km, Earth at 152,000,000 km, and Mars at 249,000,000 km. The surface area of each of the planets is 74,800,000 km², 460,270,000 km², 510,070,000 km², and 55,870,000 km², respectively. Halley's Comet's maximum orbital distance is 5,208,800,000 km from the sun. Scientists do not know the actual area of Halley's Comet's surface. Use the information to decide if the solution answers the question. Write *yes* or *no*. If the answer is *no*, write the correct answer.

1. What is the difference between the surface areas of Earth and Mercury?
   Answer: 435.27 million km2

   _____

2. How much larger is Mars's orbit than Venus's? Mercury
   Answer: 358,000,000 km

   _____
    Venus

3. Mercury's maximum orbit is what percent of Mars's maximum orbit?
   Answer: 72%
    Earth
   _____

4. How many fewer square kilometers of surface area does Mars have compared to Venus?
   Answer: 404.4 million kilometers

   _____
    Mars

5. Which is greater: the combined surface areas of Earth and Mercury or of Venus and Mars?
   Answer: Mars has the greater surface area.

   _____

PROBLEM SOLVING

Using a Double-Line Graph

Westech is a producer of microcomputers and computer software. To help them plan production for 1986, a broken-line graph was drawn to illustrate the company's production levels for 1984 and 1985. Each of the levels for 1985 was $3\frac{1}{2}$ times greater per quarter than the 1984 levels of software production for the same quarter. Computer production was $\frac{7}{8}$ of the 1984 levels for each quarter.

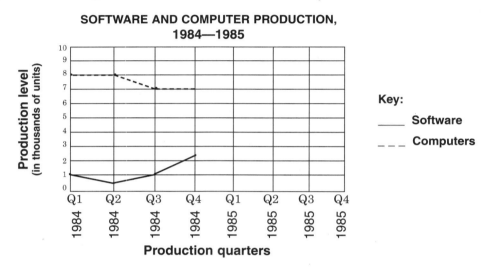

Complete the graph. Then answer the questions.

1. What was the trend for 1985 software production levels?

2. What was the 1984 computer production trend?

3. What was the best quarter for software production in 1985?

4. How does this quarter compare with the same quarter of computer production?

5. By how many units will computer production have to change in 1986 to equal 1984 levels during the first quarter?

6. For the first quarter of 1986, Westech wants their software production to be at least twice the first-quarter production level of 1985. Do they need to increase current production levels? It so, by how much?

LOGIC
Unit 3: Propositions

A **proposition** is a statement that can be shown to be true or false. Propositions are usually given letter names. This makes it easier to work with more than one proposition.

P: 1 is in square A.   Q: 1 is not in square B.

If you look at the diagram you can see that P and Q are both true propositions.

Every proposition has an opposite. The opposite of P is ~P, which is read "not P."

~P : 1 is not in square A.   ~Q : 1 is in square B.

It is easy to see that ~P and ~Q are both false propositions.

Write whether the proposition is *true* or *false* for the diagram above it.

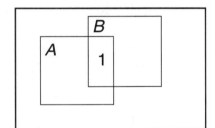 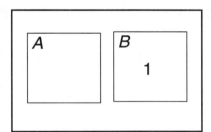 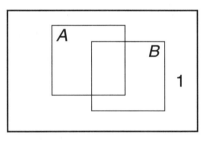

1. P _____
    ~P _____
    Q _____
    ~Q _____

2. P _____
    ~P _____
    Q _____
    ~Q _____

3. P _____
    ~P _____
    Q _____
    ~Q _____

D : 2 is not in triangle A.   R : 2 is not in triangle B.

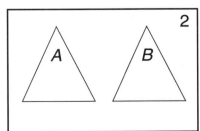 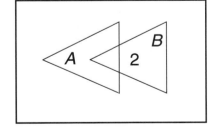 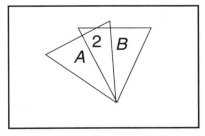

4. D _____
    ~D _____
    R _____
    ~R _____

5. D _____
    ~D _____
    R _____
    ~R _____

6. D _____
    ~D _____
    R _____
    ~R _____

Name _____ Date _____

LOGIC
Riddles

Mathematics isn't all numbers and figuring. The most important thing in mathematics is thinking. Sometimes puzzles and riddles can be good practice for mathematics.

Answer each question. Explain your answer.

1. Setsuko has two American coins in her hand. Together their value totals 55¢. One is not a nickel. What are they?

_____

2. Mr. and Mrs. Ortiz went to Mr. Davis's house for dinner. After dinner, Mr. Davis played one game of chess with each of them. He lost both games. Mr. Davis's daughter Lisa, who does not know how to play chess, said, "Dad, I'm surprised at you. I can do better than that. In fact, I'll play two games at the same time. I'll go first in the game with Mr. Ortiz, and I'll let Mrs. Ortiz go first in my game with her. I'll bet I do better than losing both games." And she did. How did she manage it?

_____
_____
_____
_____

3. The two diagrams show two different arrangements of six pennies. What is the fewest number of moves it takes to turn the arrangement on the left into the one on the right? A move consists of sliding one penny into a position where it is touching two other pennies.

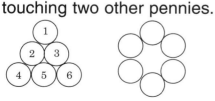

_____
_____

Name _____ Date _____

LOGIC

Logic Tables

1. Louis, Patricia, Emanuel, and Jacqueline like to swim, fly kites, read, and play chess. Each person favors one hobby the most. Find each person's favorite.

   a. No person's name has the same number of letters as his or her favorite hobby.
   b. The person who likes chess the most is a girl and is friends with both Louis and Patricia.
   c. The person who likes reading the most is a boy.

You can use a **logic table** to help you find the answers. On the table, you eliminate a box by putting an X in it, and you put an O in each box that is correct. For example, Louis's name has the same number of letters as *chess*. Statement *a* tells you that chess cannot be Louis's favorite.

	Jacqueline	Patricia	Louis	Emanuel
Chess			X	
Reading				
Swimming				
Kite flying				

2. Eyes, Ears, Nose, and Hands are secret agents. Their mission was to investigate Colonel Grabbit, who is an enemy agent. They went to his house during a party. One spy was disguised as a butler in the sitting room, one as a chef at the outdoor barbecue, and one as a coat rack in the foyer; one hid in the fountain on the lawn. Use this information to determine which was which.

   a. Eyes later said it was hard to hold still for so long.
   b. Ears went over and whispered to Hands during the party.
   c. Hands said he envied Nose and Eyes for being indoors.

	Eyes	Ears	Nose	Hands
Coat rack				
Chef				
Butler				
Fountain				

© Steck-Vaughn Company

Unit 3: Logic
Math Enrichment 6, SV8397-8

Name _____ Date _____

LOGIC

Riddles

Answer each question. Explain your answer.

1. On a remote and rather strange island in the Pacific, there are two villages. All the people in one village always tell the truth. All the people in the other village always lie. A sailor comes ashore on this island and meets two natives. He has no idea where either one of them is from. He asks the taller one, "Are you a truth-teller?" The man replies, "Dar." The sailor turns to the shorter islander and asks, "What did he say?" The man replies, "He said yes, but he's a liar." Which village is each man from?

_____
_____
_____

2. A car leaves Florida, heading for Massachusetts along Interstate 95. Another car drives from Massachusetts to Florida along Interstate 95. They start out 1,237 miles from each other. The first car drives at a constant speed of 55 miles per hour. The second car maintains a speed of 40 miles per hour. How far apart will they be one hour before they pass each other?

_____
_____
_____

3. A woman lives at the foot of a mountain. One morning, she leaves her house at 9:00 and walks up the mountain. She stops on the trail occasionally to rest or eat some fruit that she brought with her. She camps at the top, and the next morning at 9:00 she walks back down the mountain. Is there any point on the trail that she occupies at the same time on both days?

_____
_____
_____

Name _____ Date _____

LOGIC
Logic Tables

Complete each logic table. Eliminate a box by drawing an *X*, and place an *0* in each box that is correct.

1. David, Laurinda, Angela, and Lynn like to dive, ride motorcycles, fly kites, and play tennis. Among these hobbies, each person favors a different one. Use the table to find each person's favorite.

a. No person's name has the same number of letters as his or her favorite hobby.

b. The person who likes to dive the most is a girl and is friends with both Laurinda and Angela.

c. The person who likes to ride motorcycles the most has a mustache.

	Lynn	Angela	David	Laurinda
Diving				
Motorcycles				
Tennis				
Kite flying				

2. Mr. Ho, Mr. Liang, Mr. Morozumi, Mr. Lee, and Mr. Perry are cooks. At a recent food auction

a. each cook had a casserole for sale.

b. no one bought a casserole baked by her own husband.

c. Mr. Ho's casserole was bought by Mrs. Liang.

d. Mr. Lee's casserole was bought by the woman whose husband's casserole was bought by Mrs. Morozumi.

e. Mrs. Ho bought the casserole baked by the husband of the woman who bought Mr. Perry's casserole.

f. Mrs. Lee bought a casserole from someone who has her same initials.

Complete the table to indicate who bought each cook's casserole. (HINT: Think of clues in terms of unknowns. For example, clue *d*: Mrs. *X* bought Mr. Lee's casserole; Mrs. Morozumi bought Mr. *X*'s casserole.)

	Mr. Ho	Mr. Liang	Mr. Morozumi	Mr. Lee	Mr. Perry
Mrs. Ho					
Mrs. Liang					
Mrs. Morozumi					
Mrs. Lee					
Mrs. Perry					

Name _____ Date _____

LOGIC
Possible Statements

A computer's vocabulary contains only two words: ON and OFF. The computer expresses these words by means of a switch. In the diagrams at the right, Switch A is ON, and Switch B is OFF. A computer's choices can be increased by making more complicated circuits out of simple switches like A and B. In this way, computers can form "statements." In a circuit made of two switches, if the arrows represent the flow of electricity, you can see that the circuit is only on when either switch C or switch D is closed, or when both C and D are closed.

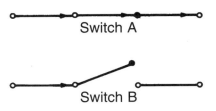

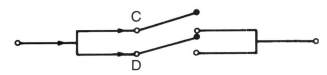

The possible statements of this circuit can be shown in a table.

C	D	Circuit
Open	Open	OFF
Open	Closed	ON
Closed	Open	ON
Closed	Closed	ON

1. Complete the table with all the possible statements for each circuit.

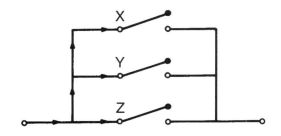

X	Y	Z	Circuit

LOGIC
Riddles

1. On a remote island in the Pacific, there are two villages. Everyone in one village always tells the truth, while everyone from the other village always lies. A sailor comes ashore and meets two islanders. He asks the taller one, "Are you a truth-teller?" The man does not speak English; so, he replies, "Dar." The sailor asks the shorter man what he said. The shorter man replies, "He said yes, but he's a liar." The sailor thinks, "A liar and a truth-teller would both answer yes to my question. *Dar* must mean "yes;" so, the short fellow must be telling the truth, and the tall one must be a liar." As it turns out, he was completely wrong. Explain.

2. Three logic students are sitting facing each other. The teacher says to them, "Close your eyes. I am going to rub chalk dust on my finger and then touch each of your foreheads. I may or may not leave a chalk mark. When you open your eyes, raise your hand if you see a mark on either of the other two students. Whichever of you can tell me for certain that you are or are not marked will receive an A." She then makes a chalk mark on all three foreheads. The students open their eyes, and all raise their hands immediately. Several moments later, student 1 says, "I know that I am marked." How did he know?

3. How can you drop an egg 14 feet without breaking it?

Unit 3: Logic

LOGIC

Logic Tables

Complete each logic table. Place an X in the boxes you eliminate and an 0 in the box with the correct answer.

1. Mr. Washington, Mr. Forrest, Mr. Perrotti, Mr. Li, and Mr. Clark are bakers. At a recent cake sale
   a. each baker had a cake for sale.
   b. no one bought a cake baked by her husband.
   c. Mr. Washington's cake was bought by Mrs. Forrest.
   d. Mr. Li's cake was bought by the woman whose husband's cake was bought by Mrs. Perrotti.
   e. Mrs. Washington bought the cake baked by the husband of the woman who bought Mr. Clark's cake.
   f. Mr. Forrest and Mr. Li made strawberry cakes. Mrs. Washington is allergic to strawberries.

Use the table to indicate who bought each man's cake.

	Mr. Washington	Mr. Forrest	Mr. Perrotti	Mr. Li	Mr. Clark
Mrs. Washington					
Mrs. Forrest					
Mrs. Perrotti					
Mrs. Li					
Mrs. Clark					

2. Five spies from different countries were spying on each other. One Thursday, they were in various spots in an art museum. In the diagram at the right, the letters refer to spaces occupied by spies. The arrows indicate which spy or spies each man could see. You also know these facts.
   a. Ax could see two other spies, while Rat could see only one.
   b. Flea was watched by only one spy, who was watched by two.
   c. Jam could be seen only by Flea.
   d. Gum was seen by both Ax and Jam.

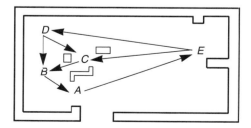

	A	B	C	D	E
Ax					
Rat					
Flea					
Jam					
Gum					

LOGIC
Possible Combinations

In the electrical-circuit diagrams at the right, current flows from A to C, as indicated by the arrows. You can see that no current will reach C unless the switch B is closed. When the electricity flows through, the circuit is ON, and when it does not flow, the circuit is OFF.

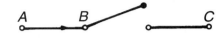

a.  b.

c. d.

If there is another switch at C, and the circuit now includes D, there are four possible combinations of open and closed switches, a-d. In only one, b, is the circuit ON.

This is another type of circuit. Draw in the switches and arrows to show the four different open/closed possibilities.

1. 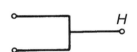 2.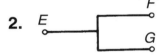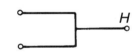

3. 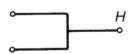 4.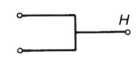

5. In Examples 1 to 4, how many show an ON circuit?

_____

In the Examples a to d, switch B *and* switch C must be closed for the circuit to be on. This is called an *and* circuit. In Examples 1 to 4, either switch F *or* switch G must be closed for the circuit to be on. This is called an *or* circuit.

6. In general, would you expect to find more ON possibilities in an *and* circuit or an *or* circuit? Explain.

_____

_____

Name _____   Date _____

LOGIC
Elimination Choices

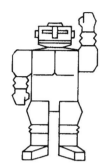

You have been hired by the We-R-Toys Company in Pittsburgh to solve a crime. At 3 P.M. on January 1, the company was vandalized. The plans to the new E-Z Robot were taken. The guilty party is one of five suspects: Fingers, Watcho, Ape, Mr. Close, and Noey. Each suspect has given four statements. Each suspect makes only one false statement. These are the statements.

WATCHO:
a. I didn't take the robot plans.
b. I've never seen We-R-Toys.
c. I've never met Mr. Close.
d. Ape lied when he said I'm guilty.

NOEY:
a. Mr. Close has used a robot.
b. The plans were taken New Year's day.
c. Ape was in New York that day.
d. One of us is guilty.

APE:
a. I was in New York when the plans were taken.
b. I've never been to We-R-Toys.
c. Watcho is the culprit.
d. Fingers is my best friend.

FINGERS:
a. Watcho has never seen We-R-Toys.
b. I don't know Ape.
c. Mr. Close was in Erie with me New Year's night.
d. I didn't take the robot plans.

MR. CLOSE:
a. I didn't take the robot plans.
b. I've never used a robot.
c. I've met Watcho before.
d. I was in Erie the night of January 1.

1. Find two statements of Noey's that you know are true.
2. Two of Watcho's statements are similar. They must be true since each suspect only makes one false statement.
3. You can find Ape's false statement from 2.

Use the table to keep track of what you deduce by writing T or F for each statement.

	a.	b.	c.	d.
Watcho				
Noey				
Ape				
Fingers				
Mr. Close				

Who is the guilty suspect? _____

PATTERNS

Unit 4: Estimating Sums

Estimate the sum of the numbers in each row of the square. Now estimate the sum of each column, then the sums of the diagonals.

60	135	30
45	75	105
120	15	90

1. What pattern do you notice? _____

Write the actual sums of the columns, rows, and diagonals.

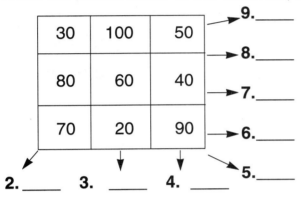

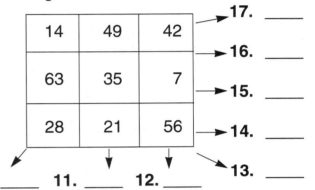

These squares are called **magic squares**. The sums are called magic sums. A **magic square** is classified by the number of boxes it has on a side. These two are *order-3* magic squares. There is an easy way to construct an order-3 square using the digits from 1 to 9.

a. Place the digits 1-9 in consecutive order.

```
1  2  3
4  5  6
7  8  9
```

b. Move each corner digit to the far side of the center digit.

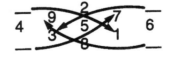

c. Shift the pattern to form a square.

```
2  7  6
9  5  1
4  3  8
```

18. What is the magic sum of this square? _____

You can construct a magic square this way using the first nine digits, counting by any number. Try it counting by 5's.

```
 5  10  15
20  25  30
35  40  45
```

19.   **20.**

21. What is the magic sum of this square? _____

47

© Steck-Vaughn Company

Unit 4: Patterns
Math Enrichment 6, SV8397-8

Name _____ Date _____

PATTERNS

Increments

Joe Farley has a job counting chickens before they're hatched. To make his job more interesting, he decided to change his **sequence of numbers**. Instead of counting 1, 2, 3, 4, and so on, he counts 1, 2.5, 4, 5.5, and so on.

1. How much greater is each number than the one before it?

_____

This number is called the **increment**.

2. What is the next number in Joe's sequence? _____

Joe's unhatched chickens are in batches of 15. After each batch, he starts counting from the beginning.

3. In his new system, what is the highest number Joe counts? _____

4. If Joe decided to start at 2, what would the first five numbers of his sequence be?

_____, _____, _____, _____, _____.

5. If Joe changed the increment to 0.5 and started at 1, what would his first five numbers be?

_____, _____, _____, _____, _____.

What would the increment be if:

6. Joe started at 1 and the third number was 5? _____

7. Joe started at 1 and the third number was 4? _____

8. Joe has labeled his unhatched-chicken-batch numbers with his 1.5 increment system. Unfortunately, he forgot to tell his supervisor about it. The supervisor sees that the last batch number is 28, but he can't find that many batches. How many batches does he actually have?

_____

9. If Joe could convince the accountant to write the amount of his check in this system, would Joe's paycheck be more or less than usual?

_____

Name _____ Date _____

PATTERNS
Arithmetic Sequences

In a **number sequence,** one number follows another in a particular order. In the sequence 2, 4, 6, 8,... the difference between any two consecutive numbers is 2. This sequence can also be written as:

2, 2 + 2, 2 + 2 + 2, 2 + 2 + 2 + 2,...

When the difference between any two consecutive numbers in a sequence is the same, it is called an **arithmetic sequence.** The difference can be any number except zero.

1. Ring the letter of each arithmetic sequence.

   a. 3.2, 6.4, 9.6, 12.8, 16, . . .   **b.** 2.5, 2.5, 2.5, 2.5,...

   c. 25, 22, 19, 16, 13,...   **d.** 6, $\frac{24}{2}$, $\frac{54}{3}$, $\frac{120}{5}$, $\frac{180}{6}$, . . .

   e. 1.3 + 1.2, 1.3 + 1.2 + 1.3, 1.3 + 1.2 + 1.3 + 1.2 . . .

2. Is the sequence 2.4, 3.6, 3.9, 5.1, 5.4,... arithmetic? Add 2.4 to 3.6, 3.6 to 3.9, 3.9 to 5.1, and 5.1 to 5.4. Write the new sequence. Is it arithmetic?

_____

Study the rows and columns of numbers.

1	2	3	4	5	6	7	8	9	10
2	4	6	8	10	12	14	16	18	20
3	6	9	12	15	18	21	24	27	30
4	8	12	16	20	24	28	32	36	40

3. Write every third number in the first row. Is this an arithmetic sequence? Where does it appear again?

_____

4. Divide every number in the third row by 3. Write the new sequence. Is it arithmetic?

_____

5. Add each column. Does any part of this arithmetic sequence appear again in the group of numbers? If so, where?

_____

_____

Name _____ Date _____

PATTERNS

Triangular and Square Numbers

Over 2,500 years ago, Greek philosophers discovered **triangular numbers** by drawing dot patterns in the sand. If the dots can be arranged as a triangle, the number is called *triangular.* 3 and 6 are triangular numbers.

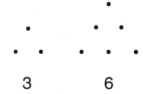

1 is also considered a triangular number. Adding successive whole numbers starting with 1 yields triangular numbers.

$1 + 2 = 3 \quad 1 + 2 + 3 = 6$

1. Bowling pins are arranged as a triangular number. With 4 rows of pins, how many pins are there? _____

2. What is the next triangular number after 10? _____

3. Draw the triangle of dots for Answer 2. _____

The same Greek philosophers discovered **square numbers**, so-called because a square number of dots can be arranged in a square pattern.

1 . , 4 :: , and 9 ::: are a few square numbers.

Adding consecutive odd numbers yields square numbers.

Show the sums as square patterns.

4. $1 + 3$ _____ **5.** $1 + 3 + 5$ _____ **6.** $1 + 3 + 5 + 7$ _____

You know that a square number can be represented by another number raised to the power of 2.

Fill in the blanks.

7. _____$^2 = 4$ **8.** $4^2 =$ _____ **9.** _____$^2 = 25$

Name _____ Date _____

PATTERNS

Fibonacci Sequence

On the planet Vinsonpreiss, the Hilpmee fly matures in 2 hours. Every hour after that, it produces, by itself, a new fly. The new fly in turn produces a fly after 2 hours. This diagram shows, if one newborn fly is put into a room, how many flies there would be after each hour. Each line represents a fly.

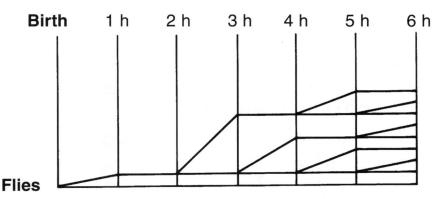

1. How many flies are there at the end of 7 hours? _____

The hourly sequence of the number of flies, 1, 1, 2, 3, 5, 8, is a special one. Each number is the sum of the previous two. This is called a **Fibonacci sequence**. It is named after a famous Italian mathematician from the thirteenth century.

Write the number of flies there will be after

2. 8 hours. _____ **3.** 9 hours. _____ **4.** 10 hours. _____ **5.** 12 hours. _____

The Roundee spider on Vinsonpreiss is similar to the Hilpmee fly, except that it produces two offspring instead of one. To find the next number in the sequence, you double the next-to-last number and add it to the last number. The sequence would start with 1, 1, 3, 5, 11, 21, and so on.

6. If there is one newborn Roundee spider at 1:00, how many will there be by 7:00? _____

7. In a room that starts with one newborn Roundee spider and one Hilpmee fly at noon, how many insects will there be by 7:00 that night? _____

© Steck-Vaughn Company

Name _____ Date _____

PATTERNS

Geometric Sequences

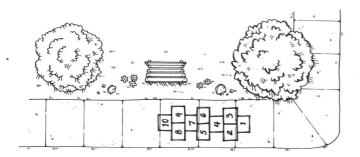

Each number in a **numeric sequence** is related to the numbers before and after it in a specific way. In the sequence 2, 4, 6, 8, 10,... each number is equal to the number before it plus 2, and the number after it minus 2. Because this sequence uses addition and subtraction to define its members, it is **arithmetic.** If members of a sequence are related by multiplication and division, the sequence is called a **geometric sequence.** The sequence 3, 9, 27, 81,... is geometric because every number in the sequence is equal to the number before it times 3, and the number after it divided by 3.

1. Ring the letter of each geometric sequence.

   a. 2, 6, 9, 14,

   b. 7, 14, 28, 56, 112,...

   c. 9,375; 1,875; 375; 75;15;...

   d. 4, 8, 12, 16, 20,...

2. Is the sequence 1, 3, 4, 12, 16,... geometric? Multiply each consecutive pair of numbers together to find four new numbers. Write the new sequence. Is it geometric?

   _____

Study the rows and columns of numbers.

2	4	8	16	32	64	128	256
4	16	64	256	1,024	4,096	16,384	65,536
8	64	512	4,096	32,768	262,144	2,097,152	16,777,216
16	256	4,096	65,536	1,048,576	16,777,216	268,435,456	4,294,967,296

3. How many geometric sequences are there? _____

4. What is similar about the second column and the top row?

   _____

   _____

5. Divide every number in the bottom row by the number above it in the third row. Write the new sequence. Is it geometric? Where does this sequence appear again?

   _____

© Steck-Vaughn Company

Name _____ Date _____

PATTERNS

Magic Square

Find the sum of each row, column, and diagonal of the square.

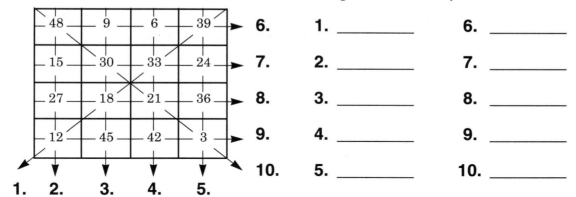

1. _____   6. _____
2. _____   7. _____
3. _____   8. _____
4. _____   9. _____
5. _____   10. _____

6. _____
7. _____
8. _____
9. _____
10. _____

A square such as this one is called a **magic square**. A magic square is classified by the number of boxes it has on a side. This is an order-4 magic square. There is an easy way to construct an order-4 square using the first 16 digits.

a. Place the digits 1 to 16 in consecutive order.

b. Exchange the ends of each diagonal.

c. Exchange the second and third digits in the middle columns.

d. Exchange the second and third digits in the end rows.

```
 1  2  3  4       16  2  3 13      16  2  3 13      16  3  2 13
 5  6  7  8        5  6  7  8       5 10 11  8       5 10 11  8
 9 10 11 12        9 10 11 12       9  6  7 12       9  6  7 12
13 14 15 16        4 14 15  1       4 14 15  1       4 15 14  1
```

You can use this method with any regular sequence of numbers.

11. Construct an order-4 magic square using these numbers.

a. 20 22 24 26
28 30 32 34
36 38 40 42
44 46 48 50

b. ___ ___ ___ ___
___ ___ ___ ___
___ ___ ___ ___
___ ___ ___ ___

c. ___ ___ ___ ___
___ ___ ___ ___
___ ___ ___ ___
___ ___ ___ ___

d. ___ ___ ___ ___
___ ___ ___ ___
___ ___ ___ ___
___ ___ ___ ___

Complete each magic square.

12.

22			10
11	20	16	
	19	15	
		18	13

13.

		7	45.5
17.5		38.5	28
31.5	21		42
14	52.5		

14.

28		35	112
	42		21
		77	14
7	84		91

Unit 4: Patterns
Math Enrichment 6, SV8397-8

Name _____ Date _____

PATTERNS
Farey Series

John Farey was a surveyor who lived during the first half of the nineteenth century. He discovered a series of fractions that have since become known as a **Farey series**. To build a Farey series, list, in order of magnitude, all the fractions between 0 and 1 that have a denominator less than or equal to a given number. For example, to build a Farey series of the number 4, list the fractions

$$\frac{1}{4}, \frac{2}{4}, \frac{3}{4}, \frac{1}{3}, \frac{2}{3}, \frac{1}{2}$$

Leave out 2/4 because it equals 1/2. Then list the fractions in ascending order.

$$\frac{1}{4}, \frac{1}{3}, \frac{1}{2}, \frac{2}{3}, \frac{3}{4}$$

Write the Farey series of each number.

1. 5: _____

2. 6: _____

3. 7: _____

One of the properties of a Farey series is that the sum of the first and last terms is always 1. The denominator of these terms is the series number.

Write the first and last terms of a Farey series of these numbers.

4. 8: _____ and _____   **5.** 11: _____ and _____   **6.** 117: _____ and _____

In the Farey series of the number 4, the neighboring fractions of $\frac{1}{3}$ are $\frac{1}{4}$ and $\frac{1}{2}$. Add the numerators of the neighboring fractions together. Then add their denominators together. The resulting fraction is $\frac{2}{6}$, which is equal to the middle fraction. This works for any fraction and its neighbors in a Farey series.

$$\frac{1+4}{4+2} = \frac{2}{6} = \frac{1}{3}$$

Write the fraction that would fit between the given fractions in a Farey series.

7. $\frac{2}{3}$ and $\frac{5}{7}$ _____   **8.** $\frac{3}{7}$ and $\frac{1}{2}$ _____   **9.** $\frac{5}{6}$ and $\frac{6}{7}$ _____

Name _____ Date _____

PATTERNS

Elements in a Set

A **set** is a group of things. In mathematics, you often use groups of numbers. This group is called Set A:

$$A = \{1,3,5,7,9,11\}$$

Each number in the set is called an element of the set. In Set A, 1 is the first number and 7 is the fourth number. *First* and *fourth* are called *ordinal* numbers because they describe the order of the set. We say that the ordinal of 7 is fourth.

1. What is the ordinal of 3? _____   **2.** What is the ordinal of 9? _____

The reverse of one of these questions might be: "What is the third element of Set A?" An easier way to write this is: "What is A_3?" Because the 3 is written below the line, it is called a *subscript*. Let's look at Set B.

$$B = \{2,4,6,8,10,12,14\}$$

3. What is B_2? _____   **4.** What is B_6? _____

Set B has seven elements. 7 describes the number of elements. It is called a *cardinal* number. We say that the *cardinality* of Set B is 7. Here are several more sets.

$C = \{3,6,9,12,15,18,21\}$ $G = \{1,47\}$
$E = \{1;11;111;1,111;11,111\}$ $Q = \{2,3,5,6,8,9\}$

5. What is the cardinality of Set G? ____   **6.** What is the ordinal of 5 in Set Q? ____

7. What is C_4? _____   **8.** What is the cardinality of Set E? ____

Look at Set C. How can we define this set without listing the elements? Each element is a multiple of 3. We can say that $C_n = n \times 3$, where $n = 1, 2, 3\ldots 7$. This means that to get Set C, we simply multiply the numbers from 1 to 7 by 3.

9. Write all the elements of Set R if $R_n = n \div 2$, where
   $n = 1, 2, 3\ldots 9$ _____.

10. What is the cardinality of Set R? _____

11. What is R_4? _____

Name _____ Date _____

PATTERNS

Magic Squares

Write the sum of the row, column, or diagonal.

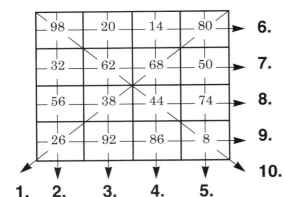

1. _____	6. _____
2. _____	7. _____
3. _____	8. _____
4. _____	9. _____
5. _____	10. _____

Squares that exhibit this property are called **magic squares**. Sums like the ones you found are called **magic sums**. Each little box in a magic square is called a **cell**.

11. Is this square magic? _____

12. What is the number in the top left cell? _____

13. Look at the number in the top left cell of the square at the top of the page. What would you multiply this number by in order to get the number in the top left cell of the second square? _____

$24\frac{1}{2}$	5	$3\frac{1}{2}$	20
8	$15\frac{1}{2}$	17	$12\frac{1}{2}$
14	$9\frac{1}{2}$	11	$18\frac{1}{2}$
$6\frac{1}{2}$	23	$21\frac{1}{2}$	2

14. Multiply the number in the bottom left cell of the square at the top of the page by 1/4. What is the product? Where else does this number appear? _____

15. Is this relationship true for every cell in the two squares?

_____

16. Is this relationship true for the magic sums of the two squares?

_____

© Steck-Vaughn Company

56

Unit 4: Patterns
Math Enrichment 6, SV8397-8

Name _____ Date _____

PATTERNS
Inverse Proportion

Two numbers are in **inverse proportion** to each other when a change in one number results in an equal and opposite change in the other number.

The speed and time it takes to travel a certain distance can be shown as an inverse proportion if the distance remains the same. You can show this as Speed × Time = Distance, or Speed = $\frac{Distance}{Time}$.

For example: 40 km/h × 2 h = 80 km, or 40 km/h = $\frac{80 \text{ km}}{2 \text{ h}}$

If you changed your speed to 50 km/h,

50 km/h × 1.6 h = 80 km, or 50 km/h = 80 km/1.6 h.

1. Kim plans to ride her bicycle 90 km. Complete the table to show how long it will take her at each different speed.

Kim's speed (km/h)	10	15	20	25	30	45
Time (h)						

The Fast Foot Shoe Store sells two kinds of designer sneakers. Sparkle sneakers cost twice as much as the solid-color sneakers. Sales of 40 pairs of solid sneakers meet the daily sales goal of $1,000.

2. What is the price per pair of each kind of sneaker?
(HINT: Use a formula like the one used for distance.
price per pair = $\frac{\text{daily sales goal}}{\text{number of pairs}}$)

Solid _____   Sparkle _____

3. How many pairs of sparkle sneakers would have to be sold to make the daily sales goal? _____

4. As the number of sales of solid sneakers went down, the store increased the price to reach the sales goal of $1,000. Complete the chart below, showing the changes in price.

Number of pairs	40	35	32	26	22	16
Price						

Name _____ Date _____

PATTERNS

Arithmetic and Geometric Sequences

If the numbers in a number sequence are related to each other by addition or subtraction, then it is an **arithmetic sequence**. If the numbers in the sequence are related by multiplication or division, it is a **geometric sequence**.

Define each sequence as *arithmetic* or *geometric*. Explain.

1. −2, 4, −8, 16, −32, ...

2. 9, 3, −3, −9,...

3. 378, 438, 498, 558, . . .

4. $21\frac{1}{3}$, 16, 12, 9, $6\frac{3}{4}$, . . .

5. 56, −2, −60, −118, −176,...

6. Does the sequence 1, 4, 10, 22, 46,... seem to be a geometric sequence? Try to multiply the first number by 2; then add 2. What is the relationship between the numbers?

7. Is 1, −1, 1, −1, 1,... a sequence? Explain.

8. Is 32, 8, 2, −1, −4, −16,... a geometric or an arithmetic sequence? If it is neither, can you rearrange the numbers so that it is one of these?

Name _____ Date _____

MEASUREMENT
Unit 5: Kilowatt-Hours

An electric bill reports usage of electricity over time. The electric company measures use of power in kilowatt-hours. A kilowatt is 1,000 watts. A **kilowatt-hour** is the number of kilowatts used in one hour. To find out how much energy is used by a 100-watt light bulb left on for 2 days, we multiply the amount of time (in hours) by the amount of energy used (in kilowatts).

2 days = 48 hours; 100 watts = 0.1 kilowatts;
48 hours × 0.1 kilowatts = 4.8 kilowatt-hours

How much energy (in kilowatt-hours) is used by

1. 2 lamps with 60-watt bulbs left on for 3 hours? _____

2. 2 lamps with 150-watt bulbs left on for $4\frac{1}{2}$ hours? _____

3. What is the combined total? _____

The electric company installs meters that show in kilowatt-hours the amount of electricity used. The meters have five dials representing ones, tens, hundreds, one-thousands, and ten-thousands.

The first dial reads 4, because the needle has not yet passed the 5. According to this meter, 47,734 kilowatt-hours have been used.

Use the list of appliance usage to calculate the kilowatt-hours logged and draw the correct meter readings.

6	100 watt	lamps	60 hours
5	60 watt	lamps	30 hours
4	150 watt	lamps	120 hours
1	1,100 watt	iron	10 hours
1	1,000 watt	vacuum	8 hours

Name _____ Date _____

MEASUREMENT

Precise Measurements

The Paddocks were planning a holiday. They planned to drive from New York City to Los Angeles. They looked on one map on which the distance was 2,451 km. A different map had a distance of 2,450 km.

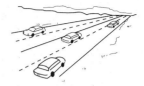

1. Which measurement is more precise? _____

Usually, a measurement that is taken to the most place values is the most precise.

Ring the measurement that is more precise.

2. 394,700 m   394,701 m   **3.** 423.4 cm   423 cm

4. 63.8273 km   63.820 km   **5.** 0.978 m   0.97 m

It will not really make a difference to the Paddocks if Los Angeles is 1 km closer or farther away. But sometimes the precision of a measurement is important.

Frank Jones makes ball bearings. His bearings have to fit very closely into holders, called *bearing races*. Because of this, his bearings must be measured very precisely, to the nearest 0.0001 cm. His bearings must be 2.3234 cm in diameter.

6. Which measurements are precise enough for Mr. Jones?

   2.3235   2.32   2.32341   3.3233   2.3233   2.3322   3.23   2.34

7. Write the measurements in order from the most precise to the least precise: 2,000 km; 2,132 km; 2,100 km; 2,130 km.

Name _____ Date _____

MEASUREMENT

Lens Magnification

Have you ever noticed that someone standing far away from you appears to be much smaller than he or she really is? You can make them seem larger by using a telescope. If the telescope can increase the apparent size of an object 7 times, it is labeled 7x; if the size increases 10 times, 10x. The x can be read as "power."

1. If a building appears 2 feet tall, how tall will it appear with a 10x telescope? _____

2. If a 300-foot tall building appears 20 feet tall, will a 10x telescope make it appear life-size? _____

There are 10 decimeters in a meter. Stretch, a basketball player, is 2 meters tall. From your seat in the stadium, Stretch appears 2 decimeters tall. Luckily, you have your 10x telescope.

3. How many decimeters tall is Stretch really? _____

2. Does your telescope make Stretch appear life-size? _____

Microscopes also use lenses to make objects appear larger. The microscope uses high powers of magnification, 100x or 1,000x or more. These are achieved by a top *eyepiece* lens and a bottom *objective* lens. To find the microscope's magnification, multiply the power of the eyepiece lens by that of the objective lens.

5. If the eyepiece lens is 10x and the objective lens is 1,000x, what is the microscope's magnification? _____ x

Complete the table.

Item to be viewed	Desired magnification	Eyepiece power	Objective power
Pollen		10	100
Fungus		30	100
Bacteria		60	100
Drop of water		20	10
Amoeba		200	100

MEASUREMENT

Determining Dates

S	M	T	W	Th	F	S
						1
2	3	4	5	6	7	8
9	10	11	12	13	14	15
16	17	18	19	20	21	22
23	24	25	26	27	28	29
30	31					

2	3	4	5	6	7
9	10	11	12	13	14
16	17	18	19	20	21
23	24	25	26	27	28
30	31				

You can tell the dates and days of the week for any year in history, or in the future, with this chart. Simply imagine moving the box across the chart to the correct day/date combination.

Leap years occur in every year divisible by 4, except century years, unless they are divisible by 400. (The year 2000 will be a leap year; the years 1700, 1800, and 1900 were not.) Ordinary years are 52 weeks and 1 day long; leap years are 1 day longer.

Presidential elections always occur on the first Tuesday in November. If January 1, 1984, was a Sunday:

1. What date was Election Day 1984? _____

2. What date was Election Day 1960? _____

Presidential elections are held every four years in the United States. (Examples 2000, 2004, 2008, 2012, 2016, 2020)

3. Are presidential elections always held in leap years?

   Explain. _____

   _____

   _____

Grover was born on the last day of February in the year 2064. He has always wanted to be President of the United States, but he must be at least 35 to be eligible.

4. What will be the first election year in which Grover will be eligible to run for the

   office of President? _____

Name _____ Date _____

MEASUREMENT

Latitude and Longitude

One of the most important people on an airplane is the navigator. The navigator's job is to chart the course of the airplane. Longitude and latitude are used for this purpose. They are imaginary lines that form a grid over the entire surface of the globe. *Latitudes* are the lines that run parallel to each other around the globe horizontally, starting at the equator (0°) and proceeding north and south to both poles (90°).

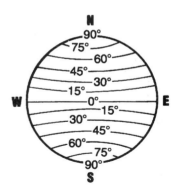

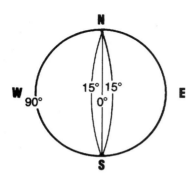

Longitudes are the lines that run from the North Pole to the South Pole, starting at the *prime meridian* (0°) and encircling the globe (0° to 180°) east and west.

1. Draw and label the longitude lines in the diagram at left.

Navigators refer to places on the globe by their *map coordinates*, which are latitude and longitude positions.

Airplane crew members can bring back presents from all over the world. Ring the places referred to by coordinates on the map, and write the correct present on the line.

	Latitude	Longitude	Presents
2.	17.3°S	149.3°W	Can of olive oil
3.	39.5°N	116.2°E	Grass skirt
4.	41.5°N	12.2°E	Panda bear
5.	4.6°N	74.1°W	Coffee beans
6.	33.6°S	151.7°E	Koala bear

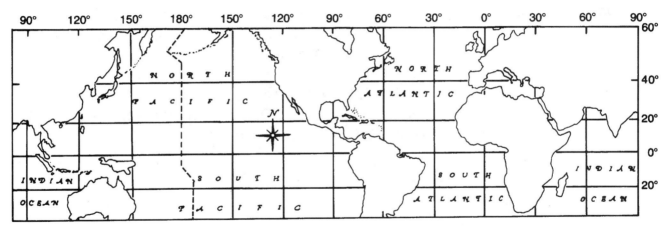

© Steck-Vaughn Company

Unit 5: Measurement
Math Enrichment 6, SV8397-8

Name _____ Date _____

MEASUREMENT

Calories

Sometimes people talk about how many **Calories** certain foods have. A Calorie (with a capital C) is actually 1,000 **calories** (with a small c). A *calorie* is the unit used to measure heat energy. 1 calorie is the amount of energy needed to raise the temperature of 1 gram of water 1° Celsius (C). Similarly, when 1 gram of water cools down 1°C, it gives off 1 calorie of heat.

1. How many calories of heat are necessary to raise 20 grams of water 20°C?

   _____

On the Celsius scale, water freezes at 0° and boils at 100°.
Room temperature is about 20°C.

2. How many calories of heat will be given off by 8 grams of water at room temperature, placed in a 0° freezer? _____

Write the answers in calories.

3. How much energy is required to boil 15 grams of ice?

   _____

4. How much energy is given off by 25 grams of boiling water that is cooling to 75°C? _____

Heat energy can be measured by a **calorimeter**. A sample of food or fuel can be burned in the device, causing the temperature of a weighed amount of water to rise. Burning a gram of sugar in the calorimeter raises the temperature of 1,000 grams of water 5°C.

5. How many calories did the sugar give off? _____

6. How many Calories were in that gram of sugar? _____

Another unit of heat, the **British Thermal Unit**, or **BTU**, is the amount of heat required to raise 1 pound of water 1° Fahrenheit (F). 1 pound of water weighs 454 grams, and a Fahrenheit degree is $\frac{5}{9}$ of a Celsius degree.

7. How many calories equal 1 BTU (to the nearest calorie)?

   _____

MEASUREMENT

Estimate Amount of Liquid

Sometimes you will find that a tool you are using to measure an object is not calibrated in small enough units for your purpose. When this happens, you may find it necessary to estimate how much you have. The beaker at the right is calibrated in ounces; so any amount of liquid that is not an exact number of ounces must be estimated. The level of liquid in this beaker is between 1 oz and 2 oz. Estimate the fraction of an ounce between 1 oz and 2 oz that the liquid occupies, and add this to the 1 oz you know you have. The level of the liquid seems to be about one third the distance between 1 oz and 2 oz; so, a good estimate of the amount of liquid in the beaker would be

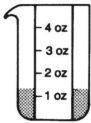

1 ounce + $\frac{1}{3}$ ounce = $1\frac{1}{3}$ ounces.

Estimate the number of ounces in each beaker.

1. 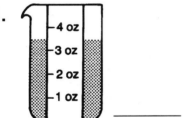 2. 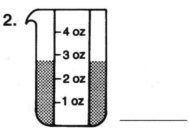 3.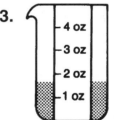

Estimate the amount of liquid to the nearest tenth of a milliliter. Use decimals.

4. 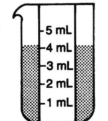 5. 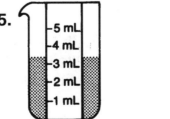 6.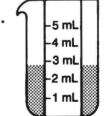

7. A 10-ounce beaker is three quarters full. If the beaker is calibrated in whole ounces, describe where the level of the liquid is.

_____

_____

MEASUREMENT

Other Units of Measure

When you measure distances, you probably use *miles*, *yards*, *feet*, and *inches*. A sailor, however, might add two other units of measurement, the *fathom* and the *league*. A horse trainer might add *furlong*. At sea, a *fathom* (6 feet) refers to the depth of the water. A *league* refers to the distance between points on the water and is 3 miles long. A horse running on a track runs distances measured in *furlongs,* 220 yards, or $\frac{1}{8}$ mile each.

1. When the men on Mississippi River steamboats measured the depth of the water as 2 fathoms, they would call "mark twain!" How many feet of water prompted this call? _____

2. There is a legend of a pair of boots that could travel 7 leagues with one step. If the circumference of Earth is 24,000 miles, how many whole steps of the 7-league boots would go around the globe? _____

3. In a battle in 1854, the Light Brigade was said to charge "half a league onward." How many furlongs is this? _____

In England, the *stone* is used as a measure of weight. 1 stone = 14 lb. Using stone (plural: *stone*) as weight measurements, answer each question.

4. How heavy are 100 boxes of cereal? _____

5. How much do you think your family car weighs? _____

There was a time when pharmacists used the *dram* as a unit of measure. One dram weighed about $\frac{1}{8}$ ounce.

6. Why do you suppose pharmacists used the dram to measure weight instead of the stone?

_____

_____

Name _____ Date _____

MEASUREMENT

Decibels

The loudness of sound is measured in **decibels**. A soft whisper measures about 10 decibels (10 dB). A noise *10 times* louder would be 20 dB. A thunderclap measures approximately 90 dB. When noises reach 140 dB, or 100,000 times louder than thunder, they become painful to the human ear.

Decibels	Noise	Decibels	Noise
10	Soft whisper	60	Noisy office
20	Quiet conversation	70	Normal traffic
30	Normal conversation	80	Rock music; subway
40	Light traffic	90	Thunder; heavy traffic
50	Ringing telephone; loud conversation	100	Jet plane taking off

1. How many times louder than a quiet conversation is a normal conversation? _____

2. How many times louder than 10 decibels is 30 decibels? _____

3. What sound is 1,000 times as loud as a quiet conversation? _____

4. Since a scream in the ear is painful, how many decibels must it measure? _____

5. How many times louder is a jet plane at takeoff than a soft whisper? _____

What is your approximation of the decibel level of the noises made by

6. a thick book hitting the floor from a height of 5 feet? _____

7. After listening to very low decibel levels, medium-range noises sound loud, and loud noises are painful. What effect do you suppose extended listening to loud noises has?

© Steck-Vaughn Company

Unit 5: Measurement
Math Enrichment 6, SV8397-8

Name _____ Date _____

MEASUREMENT

Karats

Articles made of gold are usually referred to as "14-k gold" or "16-k gold" or some other number-k gold. The *k* refers to *karats*, and it is a measurement of how much gold is in the item. A karat is $\frac{1}{24}$ the total weight of the article. Something that is pure gold is $\frac{24}{24}$, or 24-karat, gold. An item that is less than pure gold is made partly of gold, and partly of some other metal or metals, and is a fraction of 24-karat gold. A chain that is half gold and half something else is 12-karat gold.

1. What fraction of an 18-k gold watchband is actually gold? _____

2. How many karats is a chain that is $\frac{2}{3}$ gold? _____

3. If $\frac{1}{6}$ of a gold chain is a metal other than gold, how many karats is the chain? _____

4. A jeweler says you can have either a 6-ounce, 20-karat setting for your ring or an 8-ounce, 18-karat setting for the same price. Which has more gold? _____

5. Which is worth more, a 6-ounce, 12-karat gold ring or a 7-ounce, 10-karat gold ring? At a rate of $320 an ounce for gold, how much is it worth? _____

6. A 20-karat gold necklace with a number of gems on it weighs 4 pounds. If the gems weigh 2 pounds, how much gold is in the necklace? _____ pounds

7. A king has 2 gold crowns. Together they weigh 10 pounds. One crown is 20-karat gold and has the same weight of nongold in it as the second crown, which weighs 2 pounds less. How many karats is the second crown? _____

8. If the king melted down both of his crowns, how many karats would the golden lump be? _____

MEASUREMENT

Interplanetary Weights

Perhaps people from Earth will live on the different planets in our solar system in the future. While shopping in space, you'll have to remember that what weighs 1 lb on Earth will weigh more than or less than 1 lb on different planets because of their different gravitational pulls. For instance, if you wanted to buy 1 lb of green cheese on the moon, you would have the clerk weigh out only 0.16 lb. What weighs 0.16 lb on the moon weighs 1 lb on Earth.

Use the equivalence chart for interplanetary weights to solve your interplanetary shopping problems.

	Mercury	Venus	Earth	Moon	Jupiter	Saturn	Neptune
1 lb =	0.28	0.85	1.0	0.16	2.6	1.2	1.4
Diameter in miles:	3,100	7,700	7,972	2,160	88,700	75,100	27,700

1. On Venus, the apples are the ripest in the solar system. You need 5 lb (Earth weight). How much should you have the clerk weigh out for you? _____

2. Jupiter Jam is sold on Jupiter in 1-pound jars. How many jars do you need to arrive back on Earth with 10 lb of jam? _____

3. You've been asked to pick up the Saturn steak for the big barbecue on Mercury. You have to arrive on Mercury with 30 lb. How much should you buy on Saturn to make sure you will have the right amount? _____

4. If you also pick up 10 lb of pickles and 7 lb of cole slaw on Venus, how much cole slaw and pickles will you have when you arrive on Mercury? _____

5. While shopping in the Vorpal Mall on Saturn, you stop for an Omega malted. Later on in the mall, you weigh yourself on the mall scale and find that you're 3 lb heavier than before you had your malted. How many Earth pounds have you gained? _____

6. Before going back to Earth, you decide to visit the Craterama Health Spa on the moon. How many pounds will you have to lose on the moon to reach your original Earth weight? _____
How many pounds in all must you lose to come back to Earth 5 lb lighter than you left? _____

Name _____ Date _____

MEASUREMENT
....

Personal Time Line

Important moments in history and interesting dates in a person's life can be marked on a **personal time line**. The line starts near the year the person was born and ends near the person's death. The line is marked off in years. Important cultural and historic events are written below the line.

1. Calculate and write the year of each event.

- George Washington was born.   1732
- At the age of 27, Washington married Martha Custis. _____
- When Washington was 3 years old, John Peter Zenger, a New York editor, was acquitted of libel. _____
- Washington was 57 when he was elected President. _____
- Vitus Bering discovered Alaska 16 years before Mozart was born. _____
- Washington became a surveyor 10 years before he married. _____
- The War of Independence began 14 years before the French Revolution. _____

- George Washington died.   1799
- Mozart was born 3 years before Washington married. _____
- Catherine the Great became Czarina of Russia 3 years after Washington married. _____
- The French Revolution began 33 years after Mozart was born. _____
- John Adams became President 3 years before Washington died. _____
- The War of Independence began the same year Washington was elected Commander-in-Chief of the army. _____

2. Complete the time line with all the dates. List events in Washington's life above the line and cultural and historic events below the line.

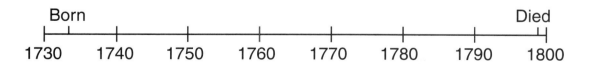

3. Write your own personal time line. Include these dates.
   1995 Yitzhak Rabin assassinated.
   1991 Persian Gulf War.
   1992 Clinton elected.
   1996 Clinton reelected.
   1994 Mandela elected president of South Africa.

Name _____ Date _____

PROBABILITY AND STATISTICS

Unit 6: Sampling

When you analyze a group of things, it is sometimes more practical to look at a small section of the group. This is called **sampling**. In order for your analysis to be accurate, the sample should be as similar as possible to the group as a whole.

Albert has just taken a job as an egg spot-checker for the Checkerspot Egg Company. Albert must check 2 cartons out of every 100 that the company ships to stores. He checks for cracked eggs, empty egg pockets, and other problems.

1. On Monday, Albert checks 2 cartons from the top of every crate of 100. Is his sampling accurate? Explain.

   _____

   _____

2. Tuesday, Albert checks 200 of the 10,000 cartons that are shipped. He finds 10 with broken eggs in them. He reports that only 10 imperfect cartons were shipped. Is his report accurate? Explain.

   _____

   _____

3. On Wednesday, Albert accidentally drops a crate of eggs. When he opens the first carton from that crate, half the eggs are broken. So are half the eggs in the second carton. He reports that half of the eggs shipped that day were broken. Is this accurate? Explain.

   _____

   _____

4. On Thursday, Albert checks 100 cartons out of 5,000 shipped. He finds only one broken egg. He reports 50 broken eggs in 5,000 cartons. Is his report accurate? Explain.

   _____

   _____

   _____

5. How could Albert's checking procedure be changed to reflect more accurately the number of irregularities found by the customer?

   _____

   _____

   _____

6. Upon opening his one thousandth carton, Albert finds a little chick chirping happily. What can Albert figure the chances to be of a chirping chick being found in a carton of eggs?

   _____

   _____

PROBABILITY AND STATISTICS

Venn Diagrams

Venn diagrams show the relationship among different sets of things. The Buncha Sums Calculator Club has 50 members. 25 have Addo machines, 10 have Neat machines, and 5 have both. Zippy machines are owned by 8 members, and 2 members own all three. The Venn diagram at the right shows this relationship. The areas where the Zippy circle intersects only the Addo and Neat circles are empty because no one has only those two machines.

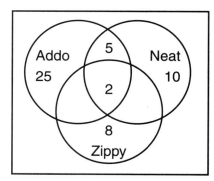

1. How many members have at least two calculators? _____

2. How many members have Addo calculators? _____

3. How many members don't have Neat calculators? _____

4. In the Always On Calculator Club, the members have the same kinds of machines as the Buncha Sums members. However, the machines are distributed differently. Use the following information to complete the chart.
    a. Twice as many members have only both Addo and Neat machines as have all three brands.
    b. 12 members have only Zippy machines.
    c. 30 members own Neat machines.
    d. 14 have both Zippy and Addo machines but do not have Neats.

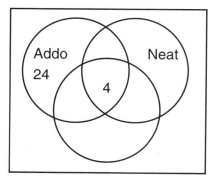

5. How many members have only both Zippy and Neat machines? _____

6. How many members own Zippy machines? _____

Name _____ Date _____

PROBABILITY AND STATISTICS

Venn Diagrams

The *Oxwatch Weekly Farm Report* found that of 200 farms, 80 have only cows, 60 have only sheep, 40 have both, and 20 have neither. A math student working in the art department decided to make a **Venn diagram** of this information, like the one at the right. The section formed by the intersecting circles represents both cows and sheep. The area outside both circles represents neither.

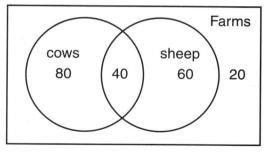

1. Fill in the diagram to show that half of the 200 farms have cows and sheep, 10 have no animals, and the number of farms that have only cows is the same as those that have only sheep.

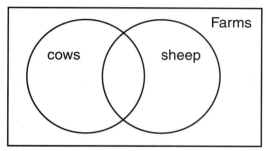  

2. How many farms have cows? _____

3. How many farms have no sheep? _____

4. How many farms have animals? _____

5. Make a Venn diagram to show this information.
   a. 175 farms have horses.
   b. 100 farms have only pigs.
   c. 75 farms have no animals, which is 50 more than have both.

6. How many farms are there altogether? _____

PROBABILITY AND STATISTICS

Probability of Genotype

Mr. and Mrs. Cleet have dark hair. Of their children, Pete, Claire, and Dot have dark hair, and Clint has red hair. The color of their hair is determined by two **genes**. A *gene* is the part of a cell that determines the characteristics or traits you inherited from your parents. Possible gene combinations can be shown in a diagram similar to a multiplication table. In the table at the right, *D* is the gene for dark hair and *r* is the gene for red. The gene for dark hair is *dominant*, which means that any pair of genes with at least one D will produce dark hair. The gene for red hair is *recessive*, meaning that both genes must be the same for the characteristic to appear. The pairs of letters within the diagram are the possible combinations, or *genotypes*, that the children could inherit.

		Mr. Cleet	
		D	r
Mrs. Cleet	D	DD	Dr
	r	Dr	rr

1. Write the genotype that Clint Cleet inherited. _____

2. Write the two possible genotypes that Claire could have inherited. _____

You can find the likelihood, or **probability**, of parents with genotype Dr having a child with dark hair. Divide the number of genotypes that produce dark hair by the total number of possible combinations. This is $\frac{3}{4}$, and means that three children of four are likely to have dark hair.

If one parent has dark hair and one parent has red hair, there are two possible gene charts. Complete the charts.

3.

	D	r
r		
r		

4.

	D	D
r		
r		

5. In Exercise 3, what is the probability of the parents having a red-haired child?

_____ of _____

6. In Exercise 4, what is the probability of the parents having a red-haired child?

_____

7. Can two red-haired parents have dark-haired children? _____

Name _____ Date _____

PROBABILITY AND STATISTICS

Batting Average

In baseball, a player's **batting average** is the number of times he hit the ball divided by the number of times he tried to hit the ball. The number is represented in thousands for greater accuracy. For example, when a batter has 50 attempts (at-bats), and gets 17 hits, his average is $\frac{17}{50}$ or .340.

1. When a batter has 12 hits in 50 at-bats, what is his batting average? _____

2. If a batter has an average of .420 after 50 at-bats, how many balls did he hit? _____

3. The same batter's average drops to .400 after 25 more at-bats. How many balls out of the 25 did he hit? _____

4. How many balls out of the 75 at-bats did he miss? _____

There are a few batters who can hit the ball from both right-handed and left-handed stances. These batters are called *switch-hitters*. Switch-hitters keep averages for both sides to see from which side they bat better. A batter might hit 30% left-handed and 35% right-handed. His total batting average is $\frac{.300 + .350}{2}$, or .325.

A switch-hitter is batting .320 right-handed after 50 at-bats. He decides to start batting left-handed. After another 50 at-bats, his overall average is .360.

5. How many hits did he get left-handed? _____

6. What was his left-handed batting average? _____

After batting 25 times from the right side and 25 times from the left, the batter has made 6 hits from the right and 8 hits from the left.

7. What was his right-handed batting average? _____

Name _____ Date _____

PROBABILITY AND STATISTICS
TV Ratings and Share

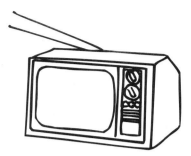

When a television show you like is canceled, it is probably because of that show's **ratings**. *Ratings* are an estimate of how many people watch a given show. They are actually a percentage shown to the first decimal place. A typical rating for a prime-time show would be 15.7. There are approximately 85.9 million households in the United States that have televisions. Because there are so many, a special sample is made by using electronic equipment. This sample involves 1,700 households.

1. What percent of all the households that have televisions takes part in the ratings sample? _____

2. If a show has a 15.7 rating, how many of the 1,700 sample households are watching? _____

There is another factor besides rating. It is called **share**. Whereas *rating* is a percentage of all the households that have televisions that are tuned in to the show, *share* is the percentage of households that have their televisions turned on. The combined rating and share would look like 15.7/30.

3. The table gives information for three separate weeks of television viewing for the 1,700 sample households for three programs. Complete the ratings and share portions. Assume that these are the only three shows on in this time-slot.

	WEEK ONE			WEEK TWO			WEEK THREE		
	Watching	Rating	Share	Watching	Rating	Share	Watching	Rating	Share
Padded Peacock	270			250			250		
Time Warp II	343			265			333		
Winnie's Wanders	211			240			257		

4. Why did *Padded Peacock* have the same share in Week One and Week Two, but have different ratings? _____

_____

5. Which show's rating increased, even though its share decreased?

_____

PROBABILITY AND STATISTICS

Venn Diagrams

Mr. and Mrs. Catino love to listen to CDs. So do their children, Matthew and Kate. There are 40 CDs in the house, of which 10 are classical music, 4 are swing music, 6 are rock and roll, 2 are nursery rhymes, 15 are folk music, and 3 are guitar lessons. Everyone likes to listen to classical music, but Matthew doesn't like swing music, the parents don't like rock and roll music, and Kate doesn't listen to folk music. Matthew is the only one who takes guitar lessons, and Kate is the only one who listens to nursery rhymes.

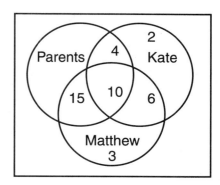

This information can be illustrated by a **Venn diagram** like the one above. Venn diagrams show how groups of things relate to each other.

1. How many CDs are only listened to by Mr. and Mrs. Catino? _____

2. How many CDs does Kate listen to, altogether? _____

3. Who listens to the most CDs? _____

The Catinos were given 28 books: 5 westerns, 9 mysteries, and 14 novels. Kate doesn't read yet, but Matthew liked 4 of the westerns, 3 mysteries, and 3 novels. Mr. Catino liked 2 of the same westerns Matthew did, and 1 Matthew didn't; 4 mysteries Matthew didn't like; and 1 novel Matthew liked and 7 Matthew didn't. Mrs. Catino didn't like any westerns, but she liked all the mysteries. She also liked all the novels, except 4 that only her husband liked.

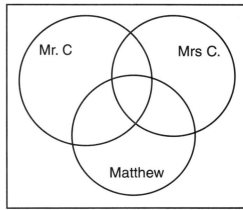

4. Use this information to fill in the Venn diagram. (HINT: Label the numbers—2M means "2 mysteries," and so on.)

5. How many books were only liked by one person? _____

Name _____ Date _____

PROBABILITY AND STATISTICS

Frequency

The most common letter in the English language is the letter E. The next most frequently used letters are T, A, O, I, R, and N. The letters used least often are Q, Z, K, X, and J.

A	B	C	D	E	F	G	H	I	J	K	L	M

N	O	P	Q	R	S	T	U	V	W	X	Y	Z

1. This is a frequency distribution chart. Fill in the number of times each letter of the alphabet is printed on this page, starting at the beginning of the first paragraph. Do not count the coded paragraph; it is not written in English.

Knowing the frequency with which letters occur can help you decipher a code. If you see a letter or number often in a coded message, it probably stands for one of the more frequently used letters. Other clues come from short, common words: a single-letter word is usually either *a* or *I*. Two-letter words are commonly *to, in, an, as, at, is,* and so on. If you have found the code for *a*, a two-letter word with *a_* is probably *an, am, as,* or *at*. A very common three-letter word is *the*. By substituting letters whose code you have broken into the coded message, you decode other letters.

2. The message is written in a code in which each letter is substituted for another letter of the alphabet. Fill in the blanks to decode the message. Some blanks have already been filled in to help you. Think about the words you know and about letters that appear next to each other in different places in words.

Utp imwe zpqp abgwchj igepigbb. G jcqb zghupn um abgw.
Utp imwe ncn hmu zghu um bpu tpq abgw, ixu utpw bpu tpq
tcu mhlp. Etp tcu g tmsp qxh. Hmz pkpqwmhp zghte tpq mh
utpcq upgs!

___ boys ____ ____ ____. __ __ wanted __ ___.

___ ___ did ___ ____ __ __ ____, but ____ let ___

hit ___. She ___ home ___. ____ everyone ____ __ __ __ __!

Name _____ Date _____

PROBABILITY AND STATISTICS

Mean

Gail needs to buy equipment for her softball team. Two stores offer different prices on the items, as shown in this chart.

	Bat	Ball	Glove
Good Sports	$15.75	$4.95	$29.99
Sportive Stuff	$16.98	$3.00	$30.19

The average, or **mean**, of a group of numbers is their sum divided by the number of numbers in the group. The mean price of an item at Good Sports is $\frac{\$15.75 + \$4.95 + \$29.99}{3}$, or $\frac{\$50.69}{3}$, or $16.90.

1. What is the mean price at Sportive Stuff? _____

It would appear that Sportive Stuff has the better buy. However, if Gail needs 4 bats, 2 balls, and 9 gloves, then the mean would have to be *weighted*, to take into account the number of times that each number in the group occurs. To obtain a *weighted mean*, multiply each number in the group by its frequency, and find the sum of these products. Divide this sum by the sum of the frequencies. For the Good Sports prices, this would be

$\frac{(4 \times \$15.75) + (2 \times \$4.95) + (9 \times \$29.99)}{4 + 2 + 9}$, or $\frac{\$63.00 + \$9.90 + + \$269.91}{15}$, or $22.85.

2. What is the weighted mean for the Sportive Stuff prices?

_____

3. For the number of articles Gail must buy, which store is the better choice?

_____

4. If Gail bought 2 bats, 10 balls, and 2 gloves, what would the weighted means for the two stores be, and which would have the better buy?

_____

_____

PROBABILITY AND STATISTICS

Odds

What do you think the odds are that 2 people in a class of 30 would have the same birthday? To calculate the probability, it is easier to ask the reverse question: What are the odds that every person in the class has a different birthday? The first person can have any birthday at all. The second person must have a birthday on one of the 364 other days in the year. The odds are equal to: $\frac{364}{365}$

The third person must have a birthday on one of the 363 days not used up by the first two people. This can be written as $\frac{364}{365} \times \frac{363}{365}$

By the time we complete this procedure for all 30 people, the expression to calculate the chances looks like this: $\frac{364}{365} \times \frac{363}{365} \times \ldots \times \frac{336}{365}$

This expression is equal to about 0.29. This means that there is a 29% chance that everyone has a different birthday. Therefore, there is a 71% chance that at least two people have the same birthday!

1. What are the chances that 2 people in a group of 20 have the same birthday? _____

2. What is the largest group possible with less than a 50% chance of 2 people having the same birthday? _____

3. What are the chances that 2 people in a group of 20, all born in a leap year, have the same birthday? _____

4. What are the chances that 2 people in a group of 20 will have birthdays in the same month? _____

5. Despite the calculations, it is difficult to believe that it is so likely that 2 people will have the same birthday. You can prove it again by experimentation. Ask each of your classmates what his or her birthday is. How many do you have to ask before you get 2 dates the same?

Name _____ Date _____

PROBABILITY AND STATISTICS

Odds

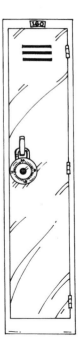

To guess a number between 1 and 20 in one try, your odds are 1 of 20, or 1/20.

If you are given 4 chances to guess the number between 1 and 20, your odds are now 4/20, or 1/5.

A friend asks you to guess 2 numbers between 1 and 3 in one guess. Assuming that the 2 numbers could be the same, here are all the combinations.

    1 and 1   2 and 1   3 and 1
    1 and 2   2 and 2   3 and 2
    1 and 3   2 and 3   3 and 3

Since there are 9 combinations of 2 numbers, you have a 1 in 9 chance of guessing the 2 numbers on the first try. There is an easier way to figure this out.

You have 3 choices for the first number and 3 choices for the second number. Multiply 3 × 3 to find the possible combinations. With only 1 guess, your odds are 1/9.

Suppose that you are asked to guess 3 numbers, each between 1 and 10.

1. What are your odds of guessing all 3 numbers on the first try? Explain.

   _____
   _____
   _____
   _____

2. You are told that the second number is 7. Now what are your odds of guessing the 3-number combination?

   _____

You arrive at school to find that the lock on your locker is not yours! It is a combination lock with 36 numbers on it. You need a 3-number combination to open the lock.

3. What are the odds of opening the lock on your first try?

   _____

PROBABILITY AND STATISTICS

Probability

A group of alphabet blocks is separated into two piles. In one pile are the letters B, D, F, H, L, J, R, O, P, V, G, T, N, Q, C, K. The other pile contains the letters I, F, M, P, A, G, W, R, E.

1. How many letters do the piles have in common? What are they? _____

2. What is the probability that a letter chosen from the first pile will be the same as a letter in the second pile? To find the answer to this question, divide the number of letters both piles have in common by the total amount of blocks in the first pile. Write the answer as a decimal.

   _____

3. In Exercise 2, you found the probability of choosing a letter from the first pile that is a letter in the second pile. Now find the probability as a decimal of choosing a letter from the second pile that will be the same as a letter in the first pile. _____

4. From which pile is there a better chance that a letter chosen will be the same as a letter in the other pile?

   _____

The following blocks are added to the first pile: I, X, Y, Z.

5. Find the new probability of choosing a letter from the first pile that is the same as a letter in the second pile.

   _____

Name _____ Date _____

ALGEBRA AND GEOMETRY

Unit 7: Binomial Expression

Any term with two numbers is called a **binomial expression**. For example, $(a+b)^2$ is a binomial expression (a and b represent any numbers you like). When you expand a binomial expression, you end up with a longer term.

Example: $(a + b)^2 = (a + b) \times (a + b)$

Think of $(a + b)$ as a 2-digit number, and expand the binomial with the same method as 2-digit multiplication.

1. Multiply $(a + b)$ by b.

$$\begin{array}{r} a + b \\ \times b \\ \hline ab + b^2 \end{array}$$ (No symbol between two letters indicates multiplication.)

2. Then multiply $(a + b)$ by a.

$$\begin{array}{r} a + b \\ \times a \\ \hline a^2 + ab \end{array}$$

3. Add the two products.

$$\begin{array}{r} a^2 + ab \\ \times ab + b^2 \\ \hline a^2 + 2ab + b^2 \end{array}$$

You have now proven that the binomial $(a + b)^2$ can be expanded into $a^2 + 2ab + b^2$. This can also be shown geometrically. Express $(a + b)^2$ as the area of a square whose sides measure $a + b$.

Mark off the lengths a and b on adjacent sides, and draw a line to the opposite side of the square. You now have four rectangles whose areas are a^2, ab, ab, and b^2. This proves geometrically that $(a + b)^2 = a^2 + 2ab + b^2$.

Substitute these numbers into your binomial-expansion formula. Show all work.

1. Let $a = 2$
   Let $b = 3$

2. Let $a = 1$
   Let $b = 2$

3. Let $a = 7$
   Let $b = 23$

4. Let $a = 5$
   Let $b = 6$

ALGEBRA AND GEOMETRY
Absolute Values

When you want to multiply positive numbers and negative numbers, the rule is to multiply the **absolute values** (the numerical amounts). When the signs are alike, the product will be positive.

Example: $(^+3) \times (^+8) = {^+24}; (^-3) \times (^-8) = {^+24}$
When the signs are not alike, the product will be negative.

Example: $(^+3) \times (^-8) = {^-24}; (^-3) \times (^+8) = {^-24}$
You can use the Distributive Property of multiplication to prove that two negative factors equal a positive product.

$$
\begin{aligned}
&= (^-2 \times {^-3}) + (0 \times {^+3}) \\
&= (^-2 \times {^-3}) + (^-2 + {^+2}) \times {^+3} \\
&= (^-2 \times {^-3}) + (^-2 \times {^+3}) + (^+2 \times {^+3}) \text{ [Distributive Property]} \\
&= {^-2} \times (^-3 + {^+3}) + (^+2 \times {^+3}) \\
&= (^-2 \times 0) + (^+2 \times {^+3}) \\
&= {^+2} \times {^+3}
\end{aligned}
$$

[Distributive Property]

Solve.

1. $^-13 \times {^-2} = $ _____
2. $^+14 \times {^-5} = $ _____
3. $^+12 \times {^-12} = $ _____
4. $^+102 \times {^-33} = $ _____
5. $^+4 \times {^+71} = $ _____
6. $^-1 \times {^+1} = $ _____
7. $^-91 \times {^+21} = $ _____
8. $^-123 \times {^-3} = $ _____
9. $^+88 \times {^-7} = $ _____
10. $^-53 \times {^+17} = $ _____

Solve.

11. $(^-3 \times {^-4}) \times (^+8 \times {^-1}) = $ _____

12. $^+2(^-2 \times {^-3}) = $ _____

13. $^+6(^+2 \times {^-3}) \times {^-2} = $ _____

14. $(^+5 \times {^-5}) \times (^-1 \times {^-4}) = $ _____

15. $(^-3 \times {^-9}) \times (^-2 \times {^+2}) \times (^-5 \times {^-2}) = $ _____

Name _____ Date _____

ALGEBRA AND GEOMETRY

Nomograph

A **nomograph** is used to determine the result of arithmetic operations on a number. The nomograph is used to find $(a - b) \div c$.

- From a point on the *a*-axis, draw a vertical line to intersect a line with a *b* value.
- Draw a horizontal line from the point of intersection to find the value of $(a - b)$.
- Continue the horizontal line to intersect a line with a *c* value.
- Draw a vertical line from the point of intersection to find the value of $(a - b) \div c$.

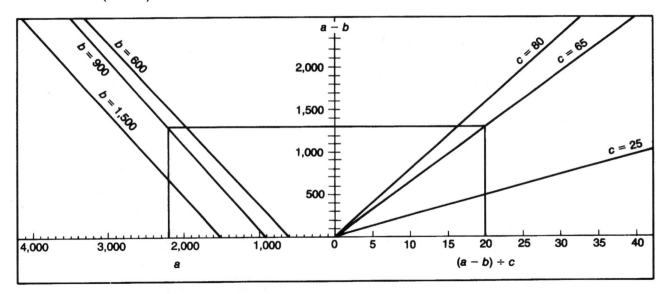

The example shown is $(2{,}200 - 900) \div 65 = 20$.
Draw lines to complete.

1. $(1{,}400 - 600) \div 25 =$ _____
2. $(2{,}700 - 1{,}500) \div 80 =$ _____

3. Create the line $c = 60$ by finding two points on the line.
   Use $0 \div 60 = 0$, and $1{,}800 \div 60 = 30$.

4. Create the line $b = 2{,}200$ by finding two points on the line.

5. $(4{,}000 - 2{,}200) \div 60 =$ _____
6. $(3{,}000 - 2{,}200) \div 80 =$ _____

© Steck-Vaughn Company

85

Unit 7: Algebra and Geometry
Math Enrichment 6, SV8397-8

Name _____ Date _____

ALGEBRA AND GEOMETRY

Networks

Network theory is an important and interesting branch of mathematics. **Networks** themselves are quite simple. They are groups of points connected by lines. The lines are called *arcs*, and the point where two or more arcs meet is called a *vertex*. (The plural of *vertex* is *vertices*.) Vertices are classified according to the number of arcs that meet at them.

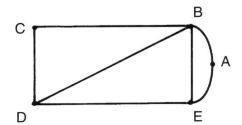
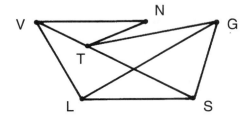

Vertex *A* is of degree 2. Vertex *S* is of degree 3.

Write the degree number of the vertex.

1. Vertex *B* _____   **2.** Vertex *C* _____   **3.** Vertex *T* _____   **4.** Vertex *L* _____

A network is said to be **traversable** if you can trace it with a continuous line that crosses each arc only once.

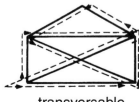
transversable

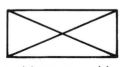
not transversable

5. Is Network *A* traversable? _____

6. How many vertices of an odd degree does it have? _____

7. How many vertices of an even degree does it have? _____

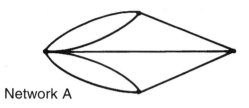
Network A

8. Is Network *B* traversable? _____

9. How many vertices of an odd degree does it have? _____

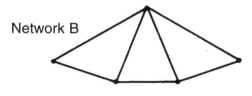
Network B

10. How many vertices of an even degree does it have? _____

© Steck-Vaughn Company

86

Unit 7: Algebra and Geometry
Math Enrichment 6, SV8397-8

ALGEBRA AND GEOMETRY

Helix

When the face of a clock is stationary, the figure made by the movement of the second hand forms a circle. Imagine the clock moving steadily upward through the air. The trail left by the second hand's motion would represent a circular spiral upward. This spiral is called a *helix*.

When the coils formed by the helix are equal in size, the helix is called *cylindrical*. When the helix becomes wider or narrower from top to bottom, it is called *conical*.

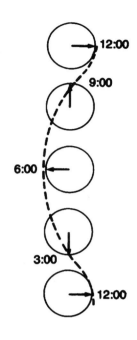

1. Moving upward at a constant rate of 12 feet per hour, how far would a clock have to travel for its minute hand to form a full turn of the helix? _____

2. What type of helix is described by the second hand of the clock described above? _____

3. What type of helix is formed by the rotor blades of a helicopter in motion? _____

CYLINDRICAL

4. What type of helix is formed by the spiraling winds of a tornado or cyclone? _____

5. If you were to compress a cylindrical helix by pressing the top and bottom toward each other, what geometric figure would you be left with? _____

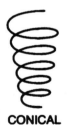

CONICAL

6. If you were to stretch out a helix from end to end, what geometric figure would you be left with? _____

7. Name five helixes in the world around you. What type of helixes do these objects form?

_____

_____

_____

Name _____ Date _____

ALGEBRA AND GEOMETRY

Fraction of Circumference

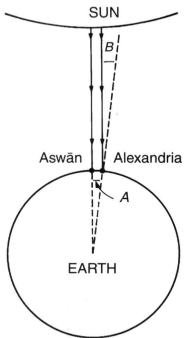

When Columbus sailed to the New World in 1492, many authorities were convinced that he would sail off the edge of the world. They thought Earth was flat, but the voyages of Columbus helped to prove that it isn't. Over 1,700 years before Columbus set sail, a Greek mathematician named **Eratosthenes** assumed that Earth was round and calculated its circumference. Eratosthenes worked as a librarian in the great library at Alexandria, Egypt. He found that at noon on a certain day, the sun was directly overhead in the city of Aswān. At the same time in Alexandria, 500 miles north of Aswān, the shadow of a pillar indicated that the sun was $7\frac{1}{2}°$ south of being directly overhead. The sun is so far away that its rays strike Earth in parallel lines. The diagram at the right shows that if angle A is $7\frac{1}{2}°$, by the laws of geometry angle B must also be $7\frac{1}{2}°$.

1. What fraction of a 360° circle is $7\frac{1}{2}°$? _____

The distance from Aswān to Alexandria is the same fraction of Earth's circumference as the answer to Exercise 1.

2. What is the circumference of Earth as Eratosthenes measured it? _____ miles

3. If the shadow of the pillar at Alexandria had indicated an angle of 9°, what would be the circumference of Earth? _____ miles

4. If the sun is directly overhead in Alexandria, Louisiana, at the same time it is 3° northwest of overhead in Texarkana, Texas, how many miles apart are these two cities? _____ miles

5. In Exercise 4, what would the circumference of Earth be if the sun were 5° northwest of overhead in Texarkana? _____ miles

6. If Eratosthenes had thought Earth was flat, how would it have affected the way he measured Earth?

Name _____ Date _____

ALGEBRA AND GEOMETRY

Pythagorean Theorem

Pythagoras was a Greek mathematician. He discovered that the length of a right triangle's hypotenuse (the side opposite the right angle), when squared, is equal to the sum of the lengths of the other two sides squared. This is called the **Pythagorean Theorem** and is expressed as $a^2 + b^2 = c^2$.

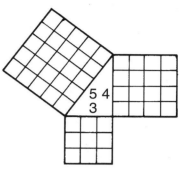

This theorem can be proved geometrically by drawing a right triangle and placing squares made up of equal sized units on each side of the triangle.

Lawanda wants to find the shortest route from her house to the store.

1. If Lawanda takes Pine Street to Oak Road to the store, how far will she have gone? _____

2. How many blocks is the route along Broadway? _____

3. Which route is shorter? _____

4. By how many blocks is it shorter?
   _____

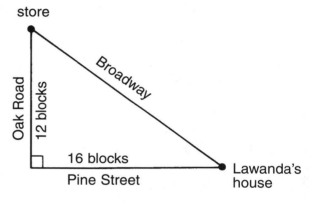

5. Jim is having a party in his backyard. He has streamers running from the top of the tree to a post in the ground. The post is 45 feet from the tree and the streamer is 54 feet long. To the nearest foot, how high is the tree?

   _____

6. Explain how this figure proves the Pythagorean Theorem.

   _____
   _____
   _____

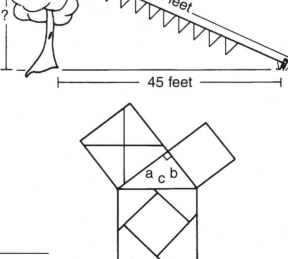

89

Name _____ Date _____

ALGEBRA AND GEOMETRY

Perimeter of Polygons

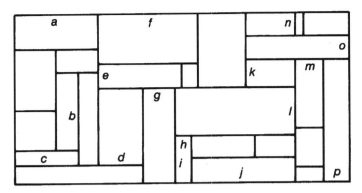

a = 11	i = 6
b = 10	j = 14
c = 8	k = 4
d = 5	l = 6
e = 4	m = 3
f = 12	n = 3
g = 4	o = 3
h = 2	p = 3

How can you find the perimeter of the large rectangle? Follow these steps.

1. What is the length of *i*? _____

2. What is the length of *i* + *l*? _____

3. What is the length of *k* + *o*? _____

4. What is the length of (*i* + *l*) + (*k* + *o*) + *n*? _____

5. What is the height of the large rectangle? _____

6. What is the length of *a* + *d*? _____

7. What is the length of *g* + *h*? _____

8. What is the length of *j* + *m*? _____

9. What is the length of (*a* + *d*) + (*g* + *h*) + (*j* + *m*) + *p*? _____

10. What is the width of the large rectangle? _____

11. What is the perimeter of the large rectangle? _____

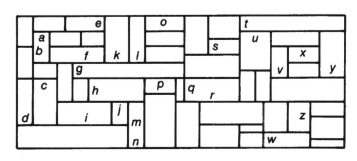

a = 2	j = 2	s = 2
b = 4	k = 3	t = 2
c = 3	l = 2	u = 4
d = 2	m = 6	v = 2
e = 2	n = 2	w = 2
f = 7	o = 5	x = 4
g = 2	p = 4	y = 3
h = 3	q = 3	z = 4
i = 7	r = 7	

12. What is the perimeter of the large rectangle? _____

Name _____ Date _____

A L G E B R A   A N D   G E O M E T R Y

Circumference

1. This wheel must travel the entire length of the line. How many turns can the wheel make before it reaches the end of the line? The radius is 12.5 inches. _____

   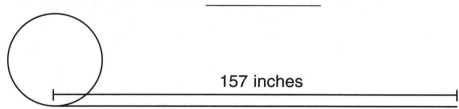
   157 inches

2. If you extended the line in Exercise 1 so that the wheel could make 5 turns, how many inches long would the line be?

   _____

3. How many inches of circumference are there between each 6-inch spoke in the diagram at right? _____

4. The circumference of the large circle is 235.5 inches. The little circle can be rolled around the large circle exactly 3 times. What is the radius of the small circle? _____

   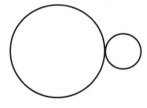

5. A ball with a diameter of 15 cm starts at point A on the floor. It rolls exactly once before it touches the wall. (REMEMBER: The part of the ball that touches the floor will not touch the wall.) What is the distance from point A to the wall? _____

6. A box that is 16.5 cm by 22 cm holds 12 balls packed tightly. What is the radius of each ball? _____

7. The diagram at the right shows a figure formed by a square between two half-circles. If the square is 4 inches on a side, how many inches around is the figure? _____

   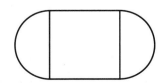

ALGEBRA AND GEOMETRY

Interior Angles

One way to find a measurement of a geometric shape is to break it down into smaller polygons.

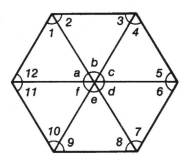

The regular hexagon has been divided into six identical triangles. Since there are 360° in a circle, the sum of the lettered angles must be 360°.

1. What is the measure of ∠b? _____

2. The sum of the interior angles of a triangle is 180°. What is the sum of ∠2 and ∠3? _____

3. All of the radii of the hexagon are of equal length; so, each triangle is isosceles. What is the measure of ∠2? _____

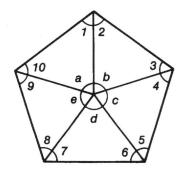

4. Since the hexagon is regular, all of the numbered angles are equal. What is the sum of the interior angles of the hexagon? _____

Use the regular pentagon to answer exercises 5 to 8.

5. What is the measure of ∠a?
_____

6. What is the sum of ∠1 and ∠10?
_____

7. What is the measure of ∠1?
_____

8. What is the sum of the interior angles of the pentagon?
_____

Complete the table.

Number of sides	Sum of interior angles
three	180°
four	
five	
six	

Name _____ Date _____

ALGEBRA AND GEOMETRY

Volume of an Irregular Solid

A **unit cube** is a cube whose edges equal one unit. Unit cubes are most useful for finding the volume of an irregular solid.

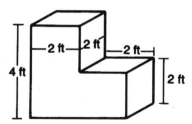

1. What is the shortest edge measurement on this irregular solid? _____

2. What is the volume of a cube with this edge measurement? _____

3. How many unit cubes like this can you fit into the irregular solid? _____

4. What is the volume of the irregular solid? _____

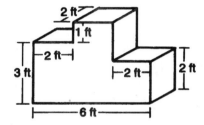

5. What is the shortest edge measurement on the solid?
_____

6. What is the volume of a unit cube with this measurement? _____

7. How many cubes like this can you fit into the solid? _____

8. What is the volume of the solid? _____

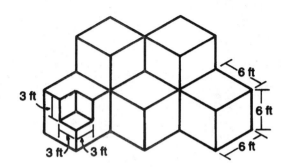

9. What is the smallest edge measurement on this irregular solid? _____

10. How many unit cubes of this measure can you fit into this solid? _____

ANSWER KEY
Math Enrichment: GRADE 6

Assessments
P. 9
1. a, 2. c, 3. b, 4. a
P. 10
1. b, 2. a, 3. c, 4. b
P. 11
1. a, 2. c, 3. d, 4. b
P. 12
1. c, 2. a, 3. a, 4. c

UNIT 1: Number
P. 13
1. 13, 2. 96, 3. 35, 4. 61, 5. 404, 6. 241, 7. 1,300, 8. 2,244
P. 14
1. 7,654 / 1,483 / +772 / 8,709 / 99 / 9,909
2. 1,732 / 71,654 / +3,994 / 75,270 / 738 / 77,380
3. 2,747 / +39 / 2,588 / 39 / 3,598
4. 661 / 1,004 / 812 / +382 / 1,947 / 20 / 2,047
5. 4,804 / 2,402 / +1,206 / 7,492 / 81 / 8,412
6. 209 / 830 / +326 / 15 / 5 / 13 / 1,365
7. 235 / 697 / +198 / 20 / 21 / 9 / 1,130
8. 378 / 816 / +22 / 16 / 10 / 11 / 1,216
9. 168 / 725 / +264 / 17 / 14 / 10 / 8 / 1,157
10. 1,333 / 7,007 / +685 / 15 / 11 / 9 / 8 / 9,025
P. 15
1. 12.50 pounds, 2. 87.5 marks, 3. $4.50; $10.00, 4. in the United States, 5. 10, 6. 5.71 francs, 7. $150, 8. 8,000 lire, 9. $10.25, 10. $8.80
P. 16
1. 25 weeks, 2. 12 weeks, 3. $11.00, 4. $8.00 each week, 5. 7 weeks, 6. She will save $3.50 less each week.
P. 17
1. 7,178, 2. 8,648, 3. 7,968, 4. 7,722, 5. 8,722, 6. 8,096, 7. 7,896, 8. 7,448, 9. 9,215, 10. 8,544, 11. 6,596, 12. 7,790, 13. 8,272, 14. 7,296, 15. 8,556, 16. 7,980, 17. 8,091, 18. 9,216, 19. The product of the two complements must be less than 100.
P. 18
1. 12,075, 2. 13,668, 3. 11,856, 4. 12,240, 5. 12,463, 6. 11,550, 7. 12,390, 8. 13,287, 9. 12,625, 10. 11,990, 11. 11,664, 12. 12,480, 13. 12,051, 14. 11,628, 15. Combine the two numbers, adding the first digit of the answer from step b to the last digit of the answer from step a.
P. 19
1. MATH BUILDS HEALTHY MINDS
2. 7-19-22-9-22 18-8 26 8-11-2 18-13
7-19-22 25-12-12-16-16-22-22-11-18-13-20 23-22-11-26-9-7-14-22-13-7
3. NO, THE SPY IS MY DOG FLUFFY.
4. Answers will vary. Possible answers include: ZIP codes, area codes, UPC codes on packages, and Dewey Decimal codes on library books.
P. 20
1.

		Those Blinker Kids	Tommy House Talk	Life's Like That!	Early Morning	Late P.M.
Cost	30	$500,000	$45,000	$20,000	$17,000	$7,500
	60	$900,000	$81,000	$36,000	$30,600	$13,500
Effective-ness	30	50	40	16	13.6	6
	60	100	80	32	27.2	12

2. ten 30-second spots on Early Morning
3. ten 30-second spots on Early Morning
4.

		Those Blinker Kids	Tommy House Talk	Life's Like That!	Early Morning	Late P.M.
Economy	30	10,000	1,125	1,250	1,250	1,250
	60	9,000	1,012.5	1,125	1,125	1,125

P. 21
1. 93,000,000 miles, 2. It would be shorter.
3. 8 minutes 19 seconds
4. 8 minutes 21 seconds
5. 8 minutes 23 seconds,
6. Answers will vary.
P. 22
1. 3; 3, 2. 1; 0, 3. 2; 2, 4. 2; 2, 5. 2; 2, 6. 6; 7, 7. 8; 7, 8. 0; 8
Ring 2, 6, 7, 8
9. Your incorrect answer might happen to have the same digit sum as the correct answer.
P. 23
1. 22, 2. 5, 3. 12, 4. 15, 5. 16, 6. 12, 7. 4, 8. 5, 9. 6
P. 24
1. $\frac{2}{9}$, 2. $\frac{7}{39}$, 3. $\frac{4}{11}$, 4. $\frac{4}{9}$, 5. $\frac{9}{10}$, 6. $\frac{1}{7}$

UNIT 2: Problem Solving
P. 25
1. Total price of prizes must be less than amount was.
2-6 Answers will vary.
P. 26

Date	Deposit	Withdrawal	Balance
Sept. 15	$30		$30
Sept. 22	$27		$57
Sept. 29	$30		$87
Nov. 3		$43	$44
Nov. 11		$10	$34
Nov. 17	$25		$59

1. $40, 2. $37, 3. $17, 4. $10, 5. yes, 6. $32
P. 27
1. Sears Building and Twin Towers—454 ft, 2. 350 ft, 3. 100 stories, 4. yes (ESB—1,472 ft), 5. 67 ft, 6. 220, 7. 4,394 ft, 8. 38 stories, 9. 1,339 ft, 10. Can't answer; information not available.
P. 28
1. $60.25, 2. $2.50, 3. $42.50, 4. $17.01 5. Total cost: $50.75; yes, 6. Answers will vary (must be $14 or under).
P. 29
1. A; 8 1/2 hours, 2. M and D; $5.63 each, 3. M, D, and A; $4.12 each, 4. Most—Mary and Phil; least—Rich and Cathy; S; $1.51, 5. A, M, S, D; more; $15.50, 6. A, M, S; more; 2.5 hours

P. 30

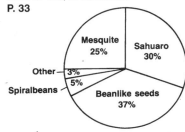

P. 31
1. 13 oblong; 1 square, 2. rectangular; 16, 3. 53 complete boxes, 4. 2 small boxes, 5. 4-15 quart and 2-5 quart, 6. 22 round tables, 7. 3-15 quart and 1-8 quart, 8. 17 boxes of forks, 9 boxes of spoons, 6 boxes of knives
P. 32
1. 77 to 69, 2. 11 to 6, 3. no, 4. yes, 5. yes; not necessarily, 6. 77 to 82, 7. 22 to 159, 8. about 283
P. 33

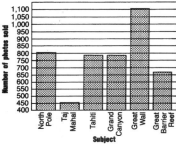

1. 12 $\frac{1}{2}$ % spiral beans, 2. 10% sahuro 3. 22.5% mesquite, 4. 5:6
P. 34

Unit 7: Algebra and Geometry

ANSWER KEY

Math Enrichment: GRADE 6

P. 34 (cont.)
1. The North Pole; 800, 2. a general increase, 3. 1st–Great Wall; 2nd–North Pole; 3rd–Tahiti and Grand Canyon (tied)

P. 35
1. yes, 2. no; 140,000,000 km, 3. no; 28%, 4. no; 404.4 million square kilometers, 5. no; Earth and Mercury combined have the greater surface areas.

P. 36

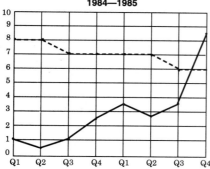

SOFTWARE AND COMPUTER PRODUCTION, 1984—1985

1. a general increase with 2nd-quarter drops in 1985, 2. a general decrease in production levels, 3. the 4th quarter 4. It is 2.625 (about 2.5) units higher for software and is increasing., 5. About a 2000-unit increase is necessary., 6. no

UNIT 3: Logic
P. 37
1. true; false; false; true
2. false; true; false; true
3. false; true; true; false
4. true; false; false; true
5. true; false; true; false
6. false; true; true; false

P. 38
1. One of the coins is not a nickel. It is a half-dollar. The other coin is a nickel.
2. Lisa waited for Mrs. Ortiz to make her first move; then she made that same move in her game with Mr. Ortiz. When Mr. Ortiz made a responding move, Lisa responded to Mrs. Ortiz with the same move. The same game was being played on both boards, with Lisa playing a different side in each. She either won one and lost one, or tied both.
3. The smallest number of moves is four: move 1 to where it touches 2 and 4; move 4 to where it touches 5 and 6; move 5 to where it touches 1 and 2; move 1 to where it touches 5 and 4.

P. 39
1.
	Jacqueline	Patricia	Louis	Emanuel
Chess	O	X	X	X
Reading	X	X	O	X
Swimming	X	X	X	O
Kite flying	X	O	X	X

2.
	Eyes	Ears	Nose	Hands
Coat rack	O	X	X	X
Chef	X	O	X	X
Butler	X	X	O	X
Fountain	X	X	X	O

P. 40
1. When asked "Are you a truth-teller?," a truth-teller will say yes and a liar will say yes. Dar must mean yes. The short man said that his friend said yes; so, he tells the truth. He also said that the tall man is a liar; so that must be true, too.
2. One hour before they meet, the first car will be 55 miles south of the passing point. The second car will be 40 miles north of the point. They will be 95 miles apart.
3. Imagine that someone duplicates her walk down the mountain, leaving at 9:00 a.m. on the day that she walks up the mountain. They must meet somewhere, so, the answer is yes.

P. 41
1.
	Lynn	Angela	David	Laurinda
Diving	O	X	X	X
Motorcycles	X	X	O	X
Tennis	X	X	X	O
Kite flying	X	O	X	X

2.
	Mr. Ho	Mr. Liang	Mr. Morozumi	Mr. Lee	Mr. Perry
Mrs. Ho	X	X	O	X	X
Mrs. Liang	O	X	X	X	X
Mrs. Morozumi	X	X	X	X	O
Mrs. Lee	X	O	X	X	X
Mrs. Perry	X	X	X	O	X

P. 42
X	Y	Z	Circuit
Open	Open	Open	OFF
Open	Open	Closed	ON
Open	Closed	Open	ON
Open	Closed	Closed	ON
Closed	Open	Open	ON
Closed	Open	Closed	ON
Closed	Closed	Open	ON
Closed	Closed	Closed	ON

P. 43
1. The tall native does not speak English; so, he would not have understood the sailor's question. *Dar* must mean "hello" or "welcome" or something other than "yes." The short man is a liar. He said his friend was a liar; so, the tall man must tell the truth.
2. Student 1 said to himself, "Let's assume that I am not marked. Student 2 knows that student 3 saw a mark because he raised his hand. 2 also knows that I am not marked. She can figure out that 3 saw a mark on her forehead. However, 2 did not say that she knew she was marked. Therefore, I must be marked."
3. Drop it from a height of 15 feet. It will drop the first 14 feet without breaking.

P. 44
1.
	Mr. Washington	Mr. Forest	Mr. Perrotti	Mr. Li	Mr. Clark
Mrs. Washington	X	X	O	X	X
Mrs. Forrest	O	X	X	X	X
Mrs. Perrotti	X	X	X	X	O
Mrs. Li	X	O	X	X	X
Mrs. Clark	X	X	X	O	X

2.
	A	B	C	D	E
Ax	X	X	X	O	X
Rat	X	O	X	X	X
Flea	O	X	X	X	X
Jam	X	X	X	X	O
Gum	X	X	O	X	X

P. 45

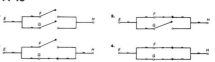

5. 3, 6. an *or* circuit; the *and* circuit requires all switches closed for ON.

P. 46
	T	T	F	T
Watcho	T	T	F	T
Noey	F	T	T	T
Ape	T	T	F	T
Fingers	T	F	T	T
Mr. Close	F	T	T	T

Who is the guilty suspect? Mr. Close

UNIT 4: Patterns
P. 47
1. They are all equal or nearly equal.
2. 180, 3. 180, 4. 180, 5. 180, 6. 180
7. 180, 8. 180, 9. 180
10. 105, 11. 105, 12. 105, 13. 105,
14. 105, 15. 105, 16. 105, 17. 105
18. 15
19.   20   45 10 35   30
           15   25   5
               40
20.   10   35   30
      45   25   5
      20   15   40
21. 75

P. 48
1. 1.5, 2. 7, 3. 22, 4. 2, 3.5, 5, 6.5, 8
5. 1, 1.5, 2, 2.5, 3, 6. 2, 7. 1.5, 8. 19
9. more

P. 49
1. a, c, d, 2. no; 6, 7.5, 9. 10.5; yes
3. 3, 6, 9; yes; in the third column
4. 1, 2, 3, 4, 5, 6, 7, 8, 9, 10; yes
5. 10, 20, 30, 40, 50, 60, 70, 80, 90, 100; yes; in the last column

P. 50
1. 10, 2. 15,
3-6. Students will draw dot patterns.
7. 2, 8. 16, 9. 5

P. 51
1. 13, 2. 21, 3. 34, 4. 55, 5. 144, 6. 43 Roundee spiders, 7. 98 (85 spiders and 13 flies)

Math Enrichment: GRADE 6

P. 52
1. b, c, 2. no; 3, 12, 48, 192; yes, 3. 12
4. The second column appears as every other number in the top row., 5. 2, 4, 8, 16, 32, 64, 128, 256; yes; in the top row

P. 53
1. 102, 2. 102, 3. 102, 4. 102, 5. 102
6. 102, 7. 102, 8. 102, 9. 102, 10. 102
11. b.

50	22	24	44
28	30	32	34
36	38	40	42
26	46	48	20

c.

50	22	24	44
28	38	40	34
36	30	32	42
26	46	48	20

d.

50	24	22	44
28	38	40	34
36	30	32	42
26	48	46	20

12.

22	17	21	10
11	20	16	23
12	19	15	24
25	14	18	13

13.

56	10.5	7	45.5
17.5	35	38.5	28
31.5	21	24.5	42
14	52.5	49	3.5

14.

28	63	35	112
105	42	70	21
98	49	77	14
7	84	56	91

P. 54
1. $\frac{1}{5}, \frac{1}{4}, \frac{1}{3}, \frac{2}{5}, \frac{1}{2}, \frac{3}{5}, \frac{2}{3}, \frac{3}{4}, \frac{4}{5}$
2. $\frac{1}{6}, \frac{1}{5}, \frac{1}{4}, \frac{1}{3}, \frac{2}{5}, \frac{1}{2}, \frac{3}{5}, \frac{2}{3}, \frac{3}{4}, \frac{4}{5}, \frac{5}{6}$
3. $\frac{1}{7}, \frac{1}{6}, \frac{1}{5}, \frac{1}{4}, \frac{2}{7}, \frac{1}{3}, \frac{2}{5}, \frac{3}{7}, \frac{1}{2}, \frac{4}{7}, \frac{3}{5}, \frac{2}{3}, \frac{5}{7},$
$\frac{3}{4}, \frac{4}{5}, \frac{5}{6}, \frac{6}{7}$
4. $\frac{1}{8}$; 5. $\frac{7}{11}; \frac{1}{11}; \frac{10}{11}$, 6. $\frac{1}{117}; \frac{116}{117}$
7. $\frac{7}{10}$, 8. $\frac{4}{9}$, 9. $\frac{1}{13}$

P. 55
1. second, 2. fifth, 3. 4, 4. 12, 5. 2,
6. third, 7. 12, 8. 5, 9. 0.5, 1, 1.5, 2, 2.5, 3, 3.5, 4, 4.5, 10. 9, 11. 2

P. 56
1. 212, 2. 212, 3. 212, 4. 212, 5. 212,
6. 212, 7. 212, 8. 212, 9. 212, 10. 212
11. yes, 12. 24 1/2, 13. 1/4, 14. 6 1/2; in the bottom left cell of the second magic square, 15. yes, 16. yes

P. 57
1.

| Kim's speed | 10 | 15 | 20 | 25 | 30 | 45 |
| Time | | 9 | 6 | 4.5 | 3.6 | 3 | 2 |

2. $25.00; $50.00, 3. 20
4.

| No. of pairs | 40 | 35 | 32 | 26 | 22 | 16 |
| Price | $25 | $28.57 | $31.25 | $38.46 | $45.45 | $62.50 |

P. 58
1. Geometric; each number is -2 times the one before., 2. Arithmetic; each number is 6 less than the one before., 3. Arithmetic; each number is 60 greater than the one before., 4. Geometric; each number is 3/4 of the one before., 5. Arithmetic; each number is 58 less than the one before.
6. No; each number is two greater than twice the one before it., 7. Yes, a geometric one; each number is -1 times the one before., 8. No; -1, 2, -4, 8, -16, 32,...

UNIT 5: Measurement
P. 59
1. 0.36 kilowatt-hours, 2. 1.35 kilowatt-hours, 3. 1.71 kilowatt-hours

P. 60
1. 2,451 km, 2. 394,701 m, 3. 423.4 cm, 4. 63.8273 km, 5. 0.978 m
6. 2.3235, 2.32341, 2.3233
7. 2,132 km; 2,130 km; 2,100 km; 2,000 km

P. 61
1. 20 feet, 2. no, 3. 20, 4. yes, 5. 10,000
Pollen - 1,000
Fungus - 3,000
Bacteria - 6,000
Drop of water - 200
Amoeba - 20,000

P. 62
1. November 6, 2. November 1
3. No; sometimes, election years fall on century years that are not divisible by 400
4. 2100

P. 63
1.

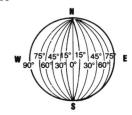

2. Grass skirt, 3. Panda bear, 4. Can of olive oil, 5. Coffee beans, 6. Koala bear
Students should accurately mark Peking, Sydney, Tahiti, Bogota, and Rome on the map.

P. 64
1. 400 calories, 2. 160 calories, 3. 1,500 calories, 4. 625 calories, 5. 5,000 calories
6. 5 Calories, 7. 252 calories

P. 65
Answers may vary. Accept any reasonable estimate.
1. $3\frac{1}{2}$, 2. $2\frac{3}{4}$, 3. $1\frac{2}{3}$, 4. 4.2 mL, 5. 3.4 mL, 6. 2.8 mL, 7. halfway between the 7-oz and the 8-oz markers

P. 66
1. 12 feet, 2. 1,143 whole steps, 3. 12 furlongs, 4. 1.5–11 stone, 5. 125–430 stone, 6. Pharmacists weigh very small amounts of medicine; so, they needed a more precise unit of measure.

P. 67
1. 10, 2. 100, 3. ringing telephone; loud conversation, 4. at least 140 dB
5. 1,000,000,000, 6. any level from 50 dB to 60 dB, 7. Medium-range noises will sound soft, and soft noises will not be heard.

P. 68
1. $\frac{3}{4}$, 2. 16, 3. 20, 4. the 8-ounce, 18-karat setting, 5. the 12-karat ring; $960, 6. 1 2/3, 7. 18, 8. 19.2

P. 69
1. 4.25 lb, 2. 26 jars, 3. 128.6 lb, 4. 5.6 lb, 5. 2.5 Earth lb, 6. 0.4 lb; 1.2 lb

P. 70
1. 1759; 1756
1735; 1762
1789; 1789
1740; 1796
1749
1775; 1775

2.

3. Answers will vary.

UNIT 6: Probability and Statistics
P. 71
1. No. He should check some from the middle or bottom, too., 2. No. The 10 he found represent 500 shipped., 3. No. His sample was based on an atypical case.
4. Yes. The one egg he found represented 50 eggs out of the whole group.
5. Checkers could go to the stores to check the eggs., 6. The chances are one in a thousand.

P. 72
1. 7, 2. 32, 3. 33
4.

[Venn diagram with three circles: Addo (24), Neat (18), Zippy (12); overlaps: Addo∩Neat = 8, Addo∩Zippy = 14, Neat∩Zippy = 4, all three = 4]

5. 0, 6. 30